Master Mind

ALSO BY DANIEL CHARLES

_____

Lords of the Harvest:
Biotech, Big Money, and the Future of Food

Master Mind

The Rise and Fall of Fritz Haber,

the Nobel Laureate Who Launched

the Age of Chemical Warfare

DANIEL CHARLES

An Imprint of HarperCollins*Publishers*

HarperCollins books may be purchased for educational, business, or sales promotional use. For information, please write: Special Markets Department, HarperCollins Publishers Inc., 10 East 53rd Street, New York, NY 10022.

The photograph showing the first railcar to leave the Leuna Works filled with ammonia in 1917 (page 258) is reproduced with the permission of the Central State Archive of Saxony-Anhalt, Merseburg Division (Landeshauptarchiv Sachsen-Anhalt, Abteilung Merseburg).

All other photographs are reproduced courtesy of the Archives for the History of the Max Planck Society in Berlin-Dahlem.

FIRST EDITION

Designed by Cassandra J. Pappas

Library of Congress Cataloging-in-Publication Data has been applied for.

ISBN 0-06-056272-2

05   06   07   08   09   WBC/RRD   10   9   8   7   6   5   4   3   2   1

To my parents

Contents

"At the end, he was forced to experience all the bitterness of being abandoned by the people of his circle, a circle that mattered very much to him, even though he recognized its dubious acts of violence. . . . It was the tragedy of the German Jew: the tragedy of unrequited love."

—Albert Einstein, in a letter about Fritz Haber

Fritz Haber in 1928, on the occasion of his sixtieth birthday.

Preface

IT'S POSSIBLE to walk in Fritz Haber's footsteps without knowing it, for the trail is rarely marked. Twenty-five years ago, as a teenager, I wandered occasionally through the courtyards of the university in Karlsruhe, Germany, where Fritz Haber first achieved fame. No statue marked his accomplishments. A few times, I rode streetcars past the hotel in Basel, Switzerland, where he died. There, too, one finds no marker.

In 1989, a month after Berlin's wall crumbled, I spent a few weeks in a disappearing country called East Germany, writing about the country's various environmental disasters. I ended up standing in the darkness on a hill near the town of Leuna, staring at distant flames that consumed exhaust gas from a chemical plant that stretched for miles across the landscape. I had no idea that this factory, the Leuna-Werke, was the fruit of Haber's labors; that it had been Germany's primary source of both munitions and fertilizer during World War I.

Eight years later, I visited the institute in Berlin that once was Fritz Haber's fiefdom. Since 1952, it has borne his name: The

Fritz Haber Institute of the Max Planck Society. I never asked who Fritz Haber was while I was there, and no one bothered to tell me. The name is mildly controversial at the institute; occasionally someone suggests that it be changed.

In the fall of 2001, the neurologist and writer Oliver Sacks waved Fritz Haber's name in front of me with a passion that I could no longer ignore. Sacks was promoting *Uncle Tungsten,* a book about his boyhood love of chemistry. This cheerful book arrived in bookstores during a dark and fearful time, immediately after the attacks of September 11 and the mysterious release of deadly anthrax spores in Florida, New York, and Washington, D.C. Naturally, an interviewer at National Public Radio asked Sacks about the ominous side of science, its power to kill and terrify. And Sacks started talking about Fritz Haber, a man who embodied the capacity of science to nourish life and destroy it.

Intrigued, I began to trace the path of Haber's life. At every turn, it led to places that were already familiar, from Karlsruhe to Berlin. And I soon realized that the legacy of this forgotten scientist was present in every day's newspaper headlines and in every bite of food.

Actually, Fritz Haber's name wasn't so much forgotten as it was driven from view. Haber died homeless in 1934, an exile from the Nazis, his name suppressed in his homeland. Later generations, both in Germany and abroad, preferred to ignore his memory.

The reason, I suspect, is that he fits no convenient category. Haber was both hero and villain; a Jew who was also a German patriot; a victim of the Nazis who was accused of war crimes himself. Unwilling to admire him, unable to condemn him, most people found it easier to look away.

Yet while Haber's name disappeared from view, the shadow of his work continued to grow.

Haber was the patron saint of guns and butter. He was a founder of the military-industrial complex and the inventor of the chemistry through which the world now feeds itself.

The Leuna Works were torn down in the early 1990s, after German unification, but a hundred factories around the world have sprouted in their place. These chemical behemoths throb with the energy of a chemical reaction first mastered by Fritz Haber, then replicated on an industrial scale by Carl Bosch.

The "Haber-Bosch process," churning out fertilizer, powers a global green revolution. It has liberated food production from the most important limits once set by nature—the limits of the soil's fertility.

The reason, in a word, is nitrogen. When we eat meat, bread, or anything containing protein, we consume nitrogen. That nitrogen comes from plants—wheat, corn, or rye—that extract it from the earth. But the earth has to be fed, too, and nature cannot on its own supply enough nitrogen to grow the food that six billion people on earth expect to eat.

Haber's process is a source of limitless nitrogen. It takes nitrogen from the air—which consists mostly of an indigestible form of nitrogen that plants can't use—and links it to hydrogen, "fixing" it in the form of ammonia. Ammonia is plant food, the most potent form of nitrogen fertilizer, though it's often converted into other forms that farmers can handle more easily.

Today, about half of all the nitrogen consumed by all the world's crops each year comes not from natural sources such as bacteria in the soil, but from ammonia factories employing the Haber-Bosch process. It has become an essential pillar of life on earth, a fountain that feeds its growing population. According to the most careful estimates, some two billion people who live on our planet today, mainly in Asia, could not survive in the absence of Fritz Haber's invention.

Hand in hand with the promise of plentiful food comes the other piece of Haber's legacy, the interweaving of science and military power. Before Fritz Haber, science and war-fighting stood at arm's length from each other. Military commanders were happy to take advantage of any technological innovation that industry or inventors might deliver—aircraft, tougher steel, steam turbines for driving ships—but the military did not run its own laboratories, and scientists did not carry new weapons into battle. As Haber once described the situation: "In the house of the German Empire, the general, the scholar, and the technologist all lived under the same roof. They greeted each other on the stairs. But there was never a fruitful exchange of ideas." The same was true in other countries.

The First World War—and Fritz Haber, the scientist-warrior—linked those worlds together more intimately than ever before. Scientists on both sides of the Atlantic created tools of warfare, but none had a greater impact than Fritz Haber. His ammonia-making process kept Germany fighting; it was Germany's main source of nitrogen, which was essential for making explosives. Haber's most notorious accomplishment, however, was the invention of a new form of warfare, as he personally directed the deployment of poisonous gas against Germany's enemies.

What began during World War I came to full fruition a few decades later, after Haber's death. Looking at pictures of Haber's institute in 1918, a fenced military compound employing hundreds of scientists and thousands of employees, one feels a premonition of the nuclear laboratory at Los Alamos, of secret cities in Russia, and scientific compounds tucked away in every other nation that lusts after the ultimate weapons of its time.

When the American military roars into battle today, airmen in supersonic jets release bombs that fly toward their targets on the wings of radio signals from orbiting satellites. Pictures of the scene arrive at my computer courtesy of the Internet, conceived and originally developed by military-funded scientists. Every part of this scene—the aircraft, the pilot, the missiles, satellites, and computer networks—represents one small piece of a scientific and military juggernaut that follows a path blazed by Fritz Haber.

Whenever Fritz Haber entered a room, his hand clenching a cigar, his bald head erect, his voice booming, he claimed the spotlight. He moved among scientific giants—Albert Einstein, Niels Bohr, Max Planck, James Franck, Max von Laue, Lise Meitner, and Otto Hahn—and he seemed the most vital force of them all.

Einstein and Bohr looked more deeply into nature's mysteries than Haber; Planck possessed more mental stamina. But none could match the speed of Haber's mental reflexes or the force of his words. Haber was an intellectual gunslinger. While others, confronted with any new idea, struggled to arrange their thoughts into logical order, Haber unleashed quick volleys of criticism and advice.

He was extravagant, impetuous, and occasionally pompous. His conversations disregarded all limits of topic or time. He loved an audience and worked out his ideas by talking aloud. But when his audience left, the internal fires subsided. In private, Haber struggled with doubts and insecurities. He was often anxious, sometimes depressed, always restless.

Lise Meitner, the codiscoverer of nuclear fission and an equally keen observer of human nature, was struck by the contrast between Haber and another prominent member of Berlin's scientific establishment, Adolf von Harnack. Harnack, she wrote, possessed an inner stability that made him seem remote and detached. Haber was the opposite, "divided within himself, and extremely passionate, which as you can imagine sometimes made things difficult for himself and for others. . . . His spontaneous reactions could be very violent and not always objective. But in the long run his generosity and reason always triumphed."

Science alone wasn't enough for Haber. He needed to do, change, create. He moved confidently between laboratory, factory, and battlefield. And wherever he went, his antennae responded to the desires of those around him. Like a boat's taut sail filled with wind, Haber absorbed the energy of his times and converted it into motion.

In any recounting of Fritz Haber's life, the Holocaust stands in the shadows, just out of sight. We know what's coming, but Fritz Haber doesn't.

With our gift of hindsight, many of Fritz Haber's passions and choices—especially his devotion to Germany—seem foolish and shortsighted. One aspect of his work seems downright macabre.

During the years immediately following World War I, Haber oversaw the research that led to the insecticide called Zyklon, then its successor, Zyklon B. A decade after his death, the SS ordered tons of that poison for the gas chambers of Auschwitz and Treblinka. Among those who died in those gas chambers were Haber's own relatives.

Despite all that, it's worth repeating the obvious: During the years when Fritz Haber climbed toward fame and fortune, the Holocaust was still unimaginable, and it was not inevitable. Haber's Germany was a nation with the same potential for good and evil as any other, unburdened by any particular load of guilt. He lived in an era of globalization, imperial rivalries, and breathtaking technological change. His life takes place, in other words, within surroundings that look surprisingly familiar to a twenty-first-century reader. And the moral choices that he confronted during his life were not so different from those that we face today.

Haber lived the life of a modern Faust, willing to serve any master who could further his passion for knowledge and progress. He was not an evil man. His defining traits—loyalty, intelligence, generosity, industry, and creativity—are as prized today as they were during his lifetime. His goals were conventional ones: to solve problems, prosper, and serve his country. And this is what makes his story tragic, for those goals, however familiar and defensible, led down twisting paths toward destruction.

Master Mind

Fritz Haber as a young boy.

ONE

Young Fritz

He was a patriot, even more extreme than I was. He was thirteen years older. The influence of time and surroundings can't be denied.

—Nobel laureate James Franck,
speaking about Fritz Haber in 1958

RAISE A BLOODY CURTAIN on the year 1871. German armies encircle and capture tens of thousands of French soldiers along the border with Belgium. Napoleon III, emperor of France, surrenders. Princes from Bavaria, Württemberg, Baden, and many other German states gather in the court of their enemy, at the palace of Versailles near Paris. They proclaim Prussia's King Wilhelm I emperor of a new German *Reich*. Germany unites; a popular dream is fulfilled. A fever of exultation sweeps the nation.

A few pessimistic voices—a tiny minority—warn of perils to come. Friedrich Nietzsche, still a young professor of classical languages, writes sourly of the "evil and perilous consequences" of wars, especially ones that have ended in victory. He complains about the delusion that German culture and civilization have triumphed, rather than simply its weapons; this delusion, he warns, was likely to lead to "the extirpation of the German spirit for the benefit of the German empire."

In the Prussian city of Breslau, a young boy bravely faces the camera, perhaps for the first time. He appears to be three or four years old. He wears his finest clothes for this portrait, and fine clothes they are. The buttons on his jacket march upward toward his neck; his hair is neatly parted and combed. He stands stiffly, left hand supporting himself on the seat of an elaborately carved chair that's just a bit taller than he is. His right hand holds the barrel of a toy gun.

The boy carries the name Fritz, the most German of names. It recalls "old Fritz," Frederick the Great, Prussian leader of the previous century.

This Fritz, however, does not appear triumphant. He looks sad and a little bit lost. His eyes are anxious. This is a picture of a motherless child.

Fritz Haber was born into a large and tightly knit Jewish clan. His parents, Paula and Siegfried, were cousins. When Paula and Siegfried were young, their families had even lived in the same house for a time, filling it with the noise and chaos of fifteen children.

Fritz was the couple's first child, arriving on December 9, 1868. It was a hard and painful birth, and Paula never recovered. She died three weeks later, on New Year's Eve.

Fritz Haber's mother, Paula, with her brothers and sisters. Paula, who died shortly after giving birth to Fritz, is second from the right among the seated women.

Siegfried, twenty-seven years old and already a successful dye merchant, was devastated. For years, he could barely face the world. He retreated into his expanding business and, according to one family member, "lived from his memories." It's unclear who cared for his infant son; one of Fritz's many aunts may have taken the boy into her home.

It was seven years before Siegfried Haber found love again. He met and married nineteen-year-old Hedwig Hamburger, noted for her beauty and her talent on the piano. Music and laughter entered the Haber house once again. And children: Three daughters—Else, Helene, and Frieda—were born within five years.

By all accounts, Hedwig became a loving stepmother to Fritz, and Fritz returned her affection. Siegfried, on the other hand, doted on his daughters; he never found it within himself to fully love or accept the son whose birth had brought so much sadness.

Fritz grew into a talkative, energetic teenager, an enthusiastic student but not a spectacularly gifted one. He soaked up everything available to an upper-middle-class boy in Breslau: theater, an education heavy in classical philosophy and literature at the elite school known as the *Gymnasium,* and hours of friendly debate and drinking in the city's beer cellars.

At home, he fought with his father. "The two of them were too different," wrote one relative. The father was cautious; the son reckless. Siegfried was a "born pessimist"; Fritz found promise in every new possibility. Siegfried regularly chased guests out the door at ten o'clock by opening windows and remarking that he "liked to air things out before everyone leaves"; Fritz, throughout his life, lost track of time when engaged in conversation. Where Siegfried was "devoid of imagination," Fritz fairly bubbled with fantastic ideas and creatively embroidered tales. Siegfried kept close account of his money; Fritz let it run through his fingers.

One younger relative was moved to wonder what Paula Haber had been like, since Fritz seemed to have inherited nothing of his father's temperament. Another, however, suggested that dour Siegfried, rather than Fritz, was the oddity in the family tree. Several of Siegfried's sisters and brothers had led lives of adventure, settling in far-off places such as Japan and the United States. And all, apart from Siegfried, had been irrepressible talkers, "a characteristic that's been inherited by Fritz Haber and his generation."

While Siegfried Haber acted as patriarch and domestic despot, Fritz became the court jester. Stories of him in this role abound. Once, when his sisters were six, four, and two years old, their mother discovered them lined up in small chairs in their room while Fritz marched back and forth in front of them, speaking loudly but incomprehensibly. Asked what he might possibly be doing, Fritz replied, "I need to acquaint my sisters with the sound of the Greek language!"

Young Fritz developed a knack for composition of rhyming verse, even on the spur of the moment. His teasing doggerel became the centerpiece of the family's annual New Year's Eve celebrations. Fritz composed the verses and taught them to his sisters, who presented them while dressed in costume. "Our childhood and youth were illuminated by our brother's talents, which always came forth at the right moment," recalled Else, the oldest of the sisters. For Siegfried, these were bittersweet occasions, for they also marked the anniversaries of Paula's death.

On the evening after Fritz graduated from the *Gymnasium* with moderately good marks in the written tests, but a masterful performance in the oral examination, he celebrated with friends in the pubs of Breslau until the wee hours of the morning. Fritz was sound asleep when the family breakfast began at seven-fifteen. Siegfried, furious, marched the three sisters to Fritz's bedroom. "Look well!" the father told them. "This is how the life of a drunkard begins!" More remarkable than the episode itself is its apparent wounding effect on the son. Forty years later, Fritz Haber's eyes filled with tears and his voice shook as he told the story to one of his closest friends.

Young Fritz wanted out. Out of his father's house, preferably out of Breslau altogether. For evidence of this, we have Fritz's own words, scribbled in letter after letter to his friend

Max Hamburger, who was learning the textile trade in distant cities.

Breslau, Fritz wrote, was an intellectual swamp. "Nothing, absolutely nothing; nothing satisfying to do, no stimulation, only irritation and tedium, having to watch out for this and that; I'm so disgusted with my entire life here that I could burst. . . .

"It's the same feeling that makes both of us dissatisfied—the urge to extricate oneself from narrow surroundings; to abandon, at all costs, the harbor into which my father has withdrawn himself after arduously weathering the storms of life; to sail out into the limitless ocean of life and future, guided by no other star than by one's own will and striving. Oh, I could work myself into a fury over the exaggerated caution that continually limits this yearning, which the mature understanding of age considers correct without recognizing that the word 'limit' is a curse that we fight against, because we only want one limit, the limit of our own ability. I hope to achieve what I want when I first break free of the chains that wear me down and confine me within circles that I find repugnant. . . . First, I'll go to university. . . ."

In Haber's mind, the university represented his longed-for port of departure. He dreamed of the "hours of genuine life" that he might experience there. And during the summer that followed his graduation from *Gymnasium*, he decided that he wanted to study chemistry. As a teenager, Haber had already begun homemade experiments, some of them dangerous. One of his aunts, who lived alone, allowed the budding chemist to use one of the rooms in her house for his experiments.

A familiar obstacle, his father, stood in the way. University studies were expensive; a year at university cost nearly as much as the total annual earnings of the best-paid mine workers of the time.

Fritz would need the financial support of his father, and Siegfried wasn't inclined to offer it.

Money wasn't the issue for Siegfried Haber. He could pay. But like all pragmatists, he readily accepted the conventions of his time and the limits imposed by circumstance. He knew exactly who he was—a Jew in Breslau—and refused to step beyond what was safe or prudent for a Jew of Breslau in the latter years of the nineteenth century. Siegfried pursued ambitions that were within reach: prosperity and a position of respect within his local community. Eventually, he would become a member of the Breslau city council.

The academic world, on the other hand, was unfamiliar and inhospitable. Siegfried had not attended university, nor had any of his ancestors. Jews were rare in the upper ranks of Germany's universities. The elder Haber foresaw roadblocks at every turn, eventual failure and frustration.

It was Siegfried's cousin Hermann, brother of his first wife Paula, who broke the deadlock. Hermann was like a brother to Siegfried; the two were the same age, and had grown up together in that busy house in the center of Breslau. Hermann was also a merchant, a wool trader. Unlike Siegfried, though, Hermann was an optimist; he was open to all modern things, wrote his granddaughter, convinced "that everything would get better." It was a reasonable expectation; at that time, everywhere one looked, yesterday's dreams—of national unity, of prosperity, of full civil rights for Jews—had become reality. Why look to the past for guidance? Why not seize the opportunities of the future? Hermann became Fritz's advocate, and Siegfried relented. In the fall of 1886, Fritz Haber left Breslau, bound for the university in Berlin.

M any years later, after Fritz Haber's death, his sister Else warned a biographer that it would be a mistake to over-emphasize her brother's Jewishness. Jewish ritual was absent from the Haber household. Yom Kippur was less important than Christmas—though tree, gifts, and meals were the center of the holiday, not the Christ child.

Religion itself was changing. Rabbis reinterpreted the scriptures, creating new forms of Judaism for a new age of reason and political freedom. They abolished many practices that made Jews distinctive. Services were held on Sunday, dietary restrictions abandoned. In a small Prussian town, Victor Klemperer's mother bought pork sausage for the first time and tasted it with reverence. It was, her son wrote later, "a form of communion," a passage into new life.

Barriers of all sorts were falling. Fritz Haber was part of the first generation of Prussian Jews in a thousand years who could imagine that their Jewish heritage might create no insurmountable hurdles in life. In 1871, the unified German *Reich* had abolished all restrictions on civil rights based on "religious difference." Some regions of Germany had dropped restrictions on Jews many years earlier. Jews became teachers, civil servants, and were elected to public office. Discrimination persisted, and anti-Jewish campaigns erupted from time to time, but for Fritz Haber, all doors seemed open. He wanted to rush through them. It was, in its own way, a response to his Jewish heritage.

F ritz Haber's yearnings—to transcend his past and explore a wider world—were also those of his nation. His departure from Breslau mirrored the German experience of the late nineteenth century. Provincial loyalties were loosening; grand ambitions were forming.

Before Haber's lifetime, "Germany" had been a vague and indeterminate idea. Before 1871, there was no country of the name. The people who lived along the banks of the Rhine, the Spree, and the Oder Rivers spoke many regional dialects and considered themselves mainly Silesians, or Prussians, or Bavarians. For the first time, they now began to think of themselves as "Germans."

Fritz was a child of the second German empire, the *Kaiserreich,* established just a few years after his birth. He breathed the intense nationalism of his time, and it seemed the most natural thing in the world.

The *Kaiserreich* was a Johnny-come-lately empire, and not really an empire at all. It controlled no significant foreign territories, at least nothing on the scale of the English and French holdings. But Germany was on a forced march into modernity. Its new steel mills were the most modern in Europe; its railroads the most efficient. In the 1880s, while Fritz Haber sat drinking beer in the pubs of Breslau, Germany was passing France as a military and economic power and closing the gap with Britain. Germans read of their nation's accomplishment with pride. The nation was feverish with a sense of limitless possibility.

Innovation brought social disruption. In 1850 only a quarter of Germans lived in urban centers. Fifty years later, half of them did. Millions of them left agricultural estates in Germany's eastern regions and migrated westward to the coal mines and steel mills that lined the Rhine River.

Economic growth, however, was erratic and unpredictable. Boom times were followed by economic collapse, like the rise and

crash of waves on a beach. Prices for many goods kept falling because of overproduction.

Henry Adams, a perceptive American visitor, chronicled the
transformation of Germany. He saw Germany first as a college
student in 1859, before its great transformation. At that time, he
was struck by the "blundering incapacity of the German for practical affairs." He saw no evidence that Germany would ever be
able to compete with France, England, or America. "Germany
had no confidence in herself, and no reason to feel it. She had no
unity, and no reason to want it. . . . [S]he was medieval by nature
and geography," Adams wrote, and he confessed that he liked it
that way.

At the end of the nineteenth century, Adams returned to Germany and barely recognized the place. "[H]ere was a Germany
new to mankind. Hamburg was almost as American as St. Louis.
In forty years the green rusticity of Düsseldorf had taken on the
sooty grime of Birmingham."

Suddenly, Germany resembled every other industrial society.
"Coal alone was felt—its stamp was the same as that of Birmingham and Pittsburgh. The Rhine produced the same power, and
the power produced the same people—the same mind—the same
impulse. . . . One great empire was ruled by one great emperor—
Coal."

Looking back on this upheaval in his later years, Fritz Haber
described it in poetic fashion: "The transformation of the agrar-
ian state into an industrial state broke over our new political world
like a force of nature, and no one protected himself against a
power that seemed only to bring unending bounty. . . . Coal
awoke from the deep, and its limitless riches rose to the light of the
working day."

Industrialists, steel-helmeted military officers, and university historians all preached the gospel of national progress and unity; so did religious leaders. The partisans of nationhood exhorted Germans to put aside narrow loyalties and unite in service to the cause of German power.

Science and technology were among the battering rams of the German assault. Germans considered mastery of science and technology to be among their great strengths, and the popular press helped nurture the cult of *Wissenschaft* and *Technik*. Any industrial success, scientific breakthrough, or new rail link was cause for celebration, but not so much because it increased the welfare of individuals; what really counted was its contribution to the strength and status of the nation.

Of all the sciences, Fritz Haber picked for himself the one that was most closely linked to Germany's rise. In the decades before Fritz was born, chemists had begun unlocking the treasures that lay within the muck left behind from the gas lamps that lined city streets; from this "coal tar" they brought forth dozens and eventually thousands of new chemicals. They deduced the chemical makeup of natural substances that imparted color to fabric and then surpassed nature, creating synthetic dyes with even more vivid hues. These discoveries gave birth to an entire industry, the first one that Germany dominated. Its triumph was imperial Germany's "greatest industrial achievement," according to historian David Landes. "In technical virtuosity and aggressive enterprise, this leap to hegemony, almost to monopoly, has no parallel."

In his rebellious optimism, his identification with the German cause, and his choice to pursue the opportunities of science, Fritz was living out the spirit of his times. Like most of his compatriots, he did this "almost unconsciously, with only an inkling of its deeper meaning," he admitted years later. He was riding a wave of success, and he saw no reason to question it. "What did we care about conflicts among political parties, their unfamiliar clamor about economic interests and social issues? Didn't the empire stand erect for all eternity, filling its natural borders? . . . We knew nothing of the worker, and we were naively bourgeois to our very bones."

Diversions and Conversion

THE CITY IN WHICH Fritz Haber arrived in 1886 was a noisy and high-spirited upstart among European cities. Berlin lacked the narrow, twisting medieval streets that filled Germany's oldest university cities, like Tübingen or Göttingen. It felt more like Chicago, proudly carrying the stamp of industry. Its streets were straight and broad, its buildings expansive, its inhabitants impatient.

People spoke of the bracing *Berliner Luft*, its air. The raw breeze sweeping down Berlin's wide boulevards carried the invigorating scent of new opportunities, of money to be made.

All over Germany, even in far-off cities of Poland and Austria, people caught a whiff of it, packed their bags, and headed toward Berlin. As Haber arrived in the German capital, the city was expanding in spectacular style. Much of it, at any one time, was either rising out of the ground or getting ripped down to make way for something new. Row upon row of six- and eight-story buildings, arrayed shoulder to shoulder like soldiers on parade, marched

Fritz Haber upon graduation from university.

outward from the city's core, enveloping what had been separate towns nearby. Construction still couldn't keep pace with demand. Tenements filled to overflowing, and new arrivals camped out in back alleys or shantytowns.

Haber came looking for freedom and intellectual adventure. He found mostly frustration and disappointment, or, as his friend Richard Willstätter put it, "seven and a half years of detours and wanderings."

The mumbled lectures of the great physicist Helmholtz produced confusion instead of inspiration. A leading professor of chemistry filled his lectures with elegant performances of chemical experiments that Haber found trivial and unchallenging.

Haber fell into a group of nine students who became an informal fraternity. The group spent most of its time in ill-tempered intellectual jousting. All the complicated ideas flying past his ears left him feeling intimidated. "God and the universe, soul and consciousness, idealism, realism! . . . I swim in a sea of formal dialectic and logic, unfortunately not with my former ease. More experienced swimmers have enticed me too often into rapids and eddies."

German students had a great deal of freedom, much more than university students in the United States. As one American visitor described it, the freedom of the German university "makes an American feel that in his own home universities with classroom recitations, with roll-call, obligatory chapel attendence, [and] daily prayer . . . the student is a mere boy . . . while the German student is treated as a full-grown man of proper responsibility in thought and action." The German student, like a free agent of learning, typically bounced from one university to the next before eventually taking the final examination required for his degree.

So it went with Fritz Haber. After one year in Berlin he moved to Heidelberg, in southwest Germany, where he liked it no better. Robert Bunsen, inventor of the Bunsen burner, nearly extinguished Haber's interest in chemistry. Bunsen was by then seventy-six years old, a forbidding and unapproachable figure whose main concern seemed to be enforcing proper laboratory procedure, rather than inspiring his students to plumb the secrets of nature. Bunsen's strict training probably was useful for Haber in the long run, but the young man chafed under it.

In letters to an old Breslau friend, Haber poured out his feelings of inadequacy. The deeper he submerged himself in his chosen field, the more insurmountable, it seemed, were the intellectual challenges that rose in front of him. He complained of "nervousness"—an intriguingly common diagnosis of the era, akin to anxiety disorders—and went on long walks to relieve it, a habit he would maintain throughout his life.

Heidelberg left one permanent mark on Haber—a long scar curving from the corner of his mouth down the left side of his chin. Later, many people assumed that the scar came from a ceremonial duel with swords—a common practice in Germany's nationalistic fraternities, where students considered such scars badges of honor and manhood. In reality, Haber acquired the scar in an ordinary fight, and he never explained exactly what the fight was over.

In 1888, Wilhelm II, grandson of England's Queen Victoria, an excitable young man with grandiose notions of himself and his country, ascended to Germany's throne. Wilhelm was just twenty-nine. He fancied himself a modern monarch, as vigorous and forward-looking as the times in which he lived. When gripped, as he sometimes was, by visions of a grander German destiny, Wilhelm II was prone to provoking other nations with aggressive speeches and military adventures. Over the next twenty years, he would fitfully and carelessly alienate Germany's most powerful neighbors, persuading France, Russia, and Great Britain to link their fortunes in an anti-German alliance.

F ritz Haber approached his twentieth birthday, and with it the prospect of military duty. For most, this meant three years in uniform. The well-to-do, however, could buy a better alternative. Any young man who'd attended university could serve a one-year term, as long as he paid all his own costs, including the cost of his horse and equipment. This was a considerable sum, much more than an ordinary worker's entire annual income, so of the young men who met the educational requirements—itself a minority— only about a third actually chose this route.

Fritz was among the privileged few. In the fall of 1888, with his father paying the bills, he joined a field artillery regiment stationed in his hometown of Breslau.

He could hardly have found a more convenient way to fulfill his military responsibilities. In Breslau, he renewed old friendships and pursued, unsuccessfully, the charming Julie Hamburger, sister of his friend Max. He also rushed from the horse stables of his regiment to philosophy classes at Breslau's university.

Haber found the details of military life—the constant orders, the "noisy desolation" of the firing range, the never-changing routine—tiresome and odious. The Prussian military's style and supreme self-confidence, however, cast a spell. For the rest of his life, Haber's personal habits—the way he walked, stood, and spoke—paid unconscious homage to military custom and discipline. And having glimpsed the military's ladder toward prestige, Haber immediately aspired to climb it. He wanted to become an officer.

There's no twenty-first-century equivalent for the social status that came with military rank in Haber's Germany. Prussian officers were the heirs of the nation's founding fathers; they, rather than civilian leaders, had unified the German *Reich*, and embodied honored German virtues of discipline and duty. The officer corps was both prestigious and exclusive: At the time Fritz Haber entered military service, almost half the officers still came from aristocratic families.

Every wealthy one-year volunteer, including Fritz Haber, could compete for the honor of an officer's commission. The goal was not to make the military a full-time career, but to become an officer in the reserves, available for national service in time of war. No matter what career one pursued, an officer's uniform was a social mark of distinction, more valuable in many circles than wealth or academic degrees.

Haber came close. He passed an initial test, demonstrating sufficient educational achievement, military aptitude, and commitment. His superior officers selected him as a candidate for election as an officer. But he was not among those elected to the club.

Haber knew the likely reason for his failure. He'd knocked at the door of a social bastion where his sort weren't welcome. At that time, no Prussian Jew had ever become a reserve officer, except in the medical corps. The rejection was a powerful reminder, if Haber needed any, that his heritage was a social handicap and that his father's cautions were grounded in reality.

He swallowed his disappointment and moved on: back to university studies in Berlin, and back to the student's agony of self-doubt and uncertainty. Now approaching the end of his studies, he still had no idea what he really wanted to do. Writing to his old friend Max Hamburger, he called himself a "lousy product," interested in much, but capable of little. Neither of the obvious career paths—in academia or in industry—inspired him.

During his final months in Berlin, he dipped into the booming field of organic chemistry—the study of carbon-based compounds like coal, or those that make up the human body. No discipline was producing a greater flood of new discoveries. Thousands of new carbon-based compounds were born each year, and a small army of chemists investigated their potential uses in industry, in the home, and even as medicines.

Haber added his small piece to the grand puzzle, with a dissertation on chemical reactions involving a smelly chemical called piperonal. A respected chemical journal published the results, but Haber found no pride in it. In a letter to his friend Max Hamburger, Haber declared the work "wretched."

"One and a half new compounds, produced like rolls at a bakery, all in the trash, and besides that a bunch of negative results . . . plus results that I can't publish at all out of fear that some competent chemist will look at them and prove that I don't know what I'm talking about. You learn to be modest."

On May 29, 1891, Haber appeared for his final examinations in chemistry, physics, and philosophy. Haber performed admirably in philosophy, adequately in his chosen field of chemistry, and poorly in physics. Asked how one would measure the electrical resistance of electrolytes—liquids containing electrically charged atoms—Haber admitted that he had no idea. It wasn't a brilliant performance, but it was enough to pass. The merchant's son could now call himself Dr. Fritz Haber.

During this final year of university studies, Haber met a fellow student named Richard Abegg, the free-spirited scion of a wealthy Berlin banker. The two became friends.

Abegg introduced Haber to the specialty within chemistry, so-called physical chemistry, in which Haber eventually would make his reputation. It was a brand-new field, and Haber hadn't encountered it before.

Physical chemistry, as the name implies, lies at the boundary of physics and chemistry—a boundary that disappears as one approaches it. It emerged from the recognition that energy—a concept from physics—forms part of every chemical reaction. Some chemical reactions require energy, whether in the form of heat, pressure, or electricity; other reactions release energy. And energy is stored in matter itself, in the chemical bonds, for instance, that bind two atoms of hydrogen to an atom of oxygen in a molecule of water. This is quite another thing, however, from the much greater energy that holds each atom together; the discovery that atoms could split and release *atomic* energy was still years in the future. Unlike organic chemists, who were busy arranging and rearranging carbon and hydrogen into new and possibly useful products, the newly formed band of physical chemists wanted to discover the mathematical laws underlying all chemical reactions.

Abegg and Haber both wrote to the leading teacher of physical chemistry in Germany, Wilhelm Ostwald, asking to join Ostwald's laboratory in Leipzig. Abegg's application was accepted; Haber's was not. The young man from Breslau had run into a professional dead end.

L eafing through the fragmentary records of Haber's early life, one searches in vain for hints of great promise. His teachers didn't seem to notice the qualities that others later praised so highly. None recognized a razor-sharp intellect or particular ability to soak up entire new fields of knowledge and immediately offer original contributions of his own.

That void is a reminder that spectacular success is always partly accidental. Haber's later triumphs were not inevitable; they were the products of opportunity and circumstance, combined with natural gifts and fierce drive.

Haber's loyal friend and epitaph writer Richard Willstätter, however, saw greatness at work even in Haber's early frustrations; Haber, in his friend's view, was too brilliant and too ambitious for his surroundings. "All of Haber's failures resulted from his inability to adjust to mediocrity. . . . Haber could never be satisfied with small assignments and small-minded objectives."

This description, written after Haber's death, was part of Willstätter's fierce battle to preserve his friend's memory, holding it above the drowning anti-Semitic tide. Yet there's truth in Willstätter's portrait. From the beginning, Fritz was searching for something grander than an occupation or profession. He dreamed of escape from humdrum life in the exalted pursuit of knowledge, but found himself trapped instead in the perplexing details of laboratory experiments. He was enthralled by a romantic notion of science, constantly disappointed by reality.

F ritz Haber emerged from university with no concrete plans. His father stepped in to fill the void. Siegfried Haber decided that Fritz could use some practical experience, and arranged appointments among his far-flung business contacts.

In rapid succession, Fritz visited three factories, spending anywhere from a few weeks to a few months at each one: a cellulose factory near Breslau, a distillery in Budapest, and a chemical plant in a part of Poland that Austria had annexed half a century earlier.

This last establishment, near the tiny village of Szczakowa, northwest of the historic city of Cracow, was the most desolate place that Fritz Haber had ever seen. A few miles away lay open fields that would become, half a century later, a brick-walled camp called Auschwitz. The village, Haber reported in a letter, consisted of little more than the factory, its buildings scattered across a "desert of sand, marsh, and fever. . . . There is nothing here, nothing at all" except monotony.

Yet in this isolated outpost, Haber was witnessing the unruly spirit of modern capitalism. The factory used a brand-new and more efficient process for making "soda" (sodium carbonate), an essential ingredient in the manufacture of soap, glass, and paper. It, and other factories like it, were upending an industry, bringing wealthy and entrenched competitors to their knees. Unconnected in any organic way to its surroundings, erected through the power of massed capital, the plant represented the shape of things to come. Haber sensed it, and was properly impressed.

Haber tried once more to gain a foothold in academia. In the fall of 1891, Haber returned to university, this time in Zurich, but stayed there only one semester. He told friends that his teacher was "authoritarian."

Half a year later, twenty-three-year-old Fritz Haber returned to Breslau, apparently resigned to life as a dye merchant. His dreams unfulfilled, his wanderings over, he was ready to learn his father's trade.

B ut first, a small interruption: In the Haber archive in Berlin there is a thin file containing a single sheet of paper, a copy of a letter dated 1955 from a woman living in what was then East Germany.

"My father owned a sanatorium for nervous ailments in the Thuringian Forest," begins the letter. "Around 1891, a patient came to us, a chemistry student named Fritz Haber." The writer of the letter explains that she was just three years old at the time. "He liked children more than most, and played with me quite a bit. My father told me later that Fritz Haber always told him, laughing about it, that I'd be his bride; he'd never marry anyone else." She still has Fritz Haber's picture, she writes. One of the other patients at the time, an artist, drew it.

This may have been the first time that Fritz Haber's "nerves"— today we might have called it anxiety—drove him to seek relief in a sanatorium. Many more of these visits would follow.

In this respect, as in so many, Haber was a man of his era. An epidemic of nervous disorders swept through Germany at the end of the nineteenth century. Men and women alike, whether middle-class or rich (the phenomenon seemed not to affect workers and peasants) collapsed under the strains and pressures of life. Their symptoms ranged from hyperactivity to impotence. Some refused to face the world and stayed in bed; others ventured forth, but found themselves overwhelmed by simple tasks, unable to concentrate, and afflicted by sudden panic attacks. As "neurasthenia" acquired official status, doctors proved increasingly willing to prescribe treatment, and government-sponsored insurance paid the bills. By 1900 there were five hundred institutions in Germany devoted to the care and healing of such ailments; no scenic mountaintop in the country seemed complete without one, and all of them were bustling with visitors, year in and year out.

Most historians who stumbled across this curious phenomenon quickly moved on to more respectable economic or political topics. The historian Joachim Radkau, however, became fascinated by it.

Radkau saw far-reaching significance in the epidemic; he felt it helped explain Germany's lurch toward an irrational and disastrous war in 1914. Germans like Haber who were seized by the jitters—and none appeared more jittery than Kaiser Wilhelm II—yearned for a cure, something that would bring single-minded purpose to their lives. They were pulled against all better judgment toward war. When World War I finally broke out in 1914, people celebrated as if experiencing release from psychic torture. The most enthusiastic spontaneous rallies took place in cities, where nervous complaints were most widespread, rather than in conservative villages of the countryside.

As for the epidemic's cause, the most obvious candidate seemed to be the tumult that arrived with the railroad, electricity, and the telegraph. Many speculated that people's nerves just couldn't cope with the pace of technological change.

The changes *were* mind-boggling and life-changing. Nothing invented since—not the mass-produced automobile, air travel, or the Internet—quite measures up to the impact of technologies that shaped life in the second half of the nineteenth century.

Over the entire course of human existence, people and their thoughts had moved across the land exactly as far and as fast as domesticated animals, whether horses, camels, or elephants, could carry them. This pace of life seemed natural and immutable, like the rising and falling of the sun. If a Russian czar died in St. Petersburg, it became a historical reality in Paris when messengers arrived on horseback.

Then came the machine, a tireless iron stallion belching smoke, rolling down rails of steel, flying across the countryside at two, three, and four times the speed of a horse-drawn carriage. Electricity within telegraph wires transmitted information at the speed of light. "Time and space are overcome," wrote one observer in 1893, voicing a common sentiment. "We fly through whole regions of the earth with the speed of the wind."

Yet as historian Joachim Radkau paged through the complaints of sanatorium patients, he found few that blamed new technology for their afflictions. They did complain about noise, and they worried about their ability to keep up with the steadily increasing pace of work, but something else seemed to be at the heart of the problem.

If anything, Radkau realized, these patients couldn't handle their apparent good fortune. They were flummoxed and overtaxed by new opportunities, surrounded by new possibilities in romantic relationships, at work, and in free time, but lacking the experience or the confidence to choose wisely among them. Progress, motion itself, was the religion of the day, so one had to move forward, but where? For fear of missing something wonderful, they rushed this way and that, never content to stay in one place, but never sure of their proper destination. They boarded trains bound for far-off lands, imagining themselves transported beyond pedestrian reality, yet found themselves deflated upon arrival.

Fritz Haber certainly fits Radkau's description. He'd always been a distractable young man, interested in many things, unable to focus on just one. He was privileged enough to dream, and explore many career paths. None, however, seemed fully satisfactory, and he had not managed to pursue any of them with success. Now he was returning to the least romantic option of all, a job in his father's business. No wonder his nerves rebelled.

Fritz Haber's life took three notable turns during 1892. In the spring, he went to work for his father, but the job lasted only half a year and ended badly. Fritz then moved to the provincial city of Jena, where he stayed for nearly two years as he took more chemistry classes and worked as a laboratory assistant. And soon after he arrived in Jena, he was baptized at St. Michael's, the leading church of the city.

No solidly documented explanations for these events survive. What's left are the stories that Fritz Haber told his friends many years later—and Haber was a notorious teller of tales, the more entertaining the better.

One friend, Rudolf Stern, called Haber's short-lived attempt to work with his father an "impossible alliance." Another wrote that Siegfried Haber came to regard his son as "a danger to the business." Two other accounts are more specific. According to one, Fritz and his father clashed over everything from Fritz's poor penmanship to the young man's conviction that synthetic chemicals would replace all natural dyes. A reckless business decision became the last straw. When cholera broke out in Hamburg, Fritz convinced his father to buy up large amounts of the chemical then used to treat the disease, chloride of lime. If cholera spread, as Fritz thought it would, they'd rake in large profits selling their stockpile. Unfortunately for Fritz, the outbreak subsided, and Siegfried Haber was stuck with chemicals no one needed.

In the other version of this story, Fritz shipped a large quantity of lime chloride to prospective buyers in Russia who then refused to accept it. "And you see, gentlemen," Haber would say after telling this tale, "that's the reason I can't stand the Russians!" In both versions, Siegfried Haber threw his son out of the business in a fury at the financial loss.

The move to Jena, then, was more like flight, the bedraggled retreat of a wounded soul. Jena offered Fritz Haber nothing particularly attractive. Its university, decidedly third-rate, boasted only one full professor in chemistry, Ludwig Knorr, and Knorr represented the part of the field that Haber had come to despise, namely organic chemistry. Yet that's where Haber landed. He became an unhappy and unpaid assistant in Knorr's laboratory, where he stayed for nearly two years. "I'm as comfortable as a drying fish," he wrote sadly in a rare letter to Max Hamburger. Yet his persistence in the laboratory represented something new. Fritz had finally made a firm decision about his future path. For better or for worse, he would pursue a career as an academic scientist with all his energy and resourcefulness.

Lastly, there's the matter of Fritz Haber's baptism. The baptismal certificate, dated November 1892, testifies to a change of social identity, one that in Germany invariably appears on all official documents alongside one's occupation and marital status. From this point onward, in the space provided for *Konfession*, or religious faith, Fritz Haber would write *evangelisch*—Protestant. Yet the switch of identity would never be complete; personal histories are not so easily erased. In the eyes of many, and often in his own, Haber remained as Jewish as ever.

Haber's decision was unusual but not rare. He was one of ten thousand or so Jews in Germany who converted to Christianity between 1890 and 1910. This, however, represented a small minority of Germany's Jewish community, which numbered between five hundred thousand and a million.

Certificates don't capture motivations, and Haber's remain unclear. He never wrote about them. Instead, once again, we have stories, some of them more trustworthy than others.

Many people at the time saw Haber's ambition at work, and felt that he'd traded his Judaism for a professor's chair. Jewish scholars who converted did, in fact, climb the academic ranks more rapidly, something that Haber understood or at least suspected. But many Jews, including Richard Willstätter, considered this a reason *not* to convert. Switching one's religious identity seemed "ill-mannered," Willstätter wrote, when accompanied by the prospect of material gain. Though Willstätter also wasn't particularly religious, he refused to have himself baptized, even when his academic mentor, himself a converted Jew, suggested that he do so.

Haber's second wife, Charlotte, who was also Jewish—and, separately, his son Hermann—wrote after Haber's death that conversion was Fritz Haber's way of breaking with his father, once and for all. As Fritz stepped inside St. Michael's Church, he symbolically turned his back on Breslau and Siegfried Haber. Charlotte Haber recalled that Siegfried Haber was deeply hurt: "Breaking away from Judaism, which is so much more than just a religious confession, seemed to him like a betrayal of one's ancestors, of one's own brothers and sisters and their fate."

Charlotte Haber also claimed that there were religious motives. Fritz, she wrote, had become friends in Jena with a young Protestant theologian, and was inspired by the Sermon on the Mount. This explanation seems dubious, for Fritz Haber rarely expressed religious sentiments or entered a church except on ceremonial occasions.

A more interesting and revealing story, however, comes from Haber's good friend and personal doctor Rudolf Stern. In the fall of 1926, the two friends rambled by car and by foot across the rocky Mediterranean coast near Monte Carlo. Haber and Stern shared a common origin in Breslau's Jewish community, so their conversations turned to Germany's rising tide of anti-Semitism, then to the "delicate question" of Haber's own conversion. As a partial explanation, Haber described the "boundless enthusiasm for Bismarck" and German unity that surrounded him in his youth. "We felt 100% German," Haber told Stern, "and no longer felt any ties to the Jewish religion." And what finally pushed him over the edge, he said, was a famous essay by historian Theodor Mommsen called "One More Word on Judaism."

This famous essay appeared when Haber was a twelve-year-old boy. Mommsen, its author, was a towering figure among German scholars, a one-man repository of knowledge about ancient Greece and Rome. But he also dove with relish into contemporary political controversies. And the essay that influenced Haber so deeply was Mommsen's contribution to an ugly alley fight that broke out among German intellectuals in 1880.

Another prominent historian and political philosopher, Heinrich von Treitschke, set off the battle with a willful provocation, a bellow of anti-Jewish fury. The "Semitic element" in Germany, raged Treitschke in a long newspaper commentary, had dragged Germany into a stinking swamp of materialism where only money mattered; Jewish newspaper writers mocked the fatherland; and "in thousands of German villages sits the Jew, busily buying up his neighbors." What's more, the plague was spreading: "Year after year, from the inexhaustible cradle of Poland, a throng of ambitious young trousers merchants pushes its way across our eastern border, intent on conquering Germany's stock exchanges and newsrooms with their children and their children's children." And thus the slogan, one that Treitschke claimed to hear sounding in unison from all corners of German society: The Jews are our misfortune.

Treitschke's incendiary words appeared first in Breslau's leading newspaper. It's not hard to imagine the shock around dinner tables of the Haber clan, for Treitschke's voice resounded like no other in German society. No ordinary academic figure, Treitschke was on his way to becoming the most influential political lecturer in Germany. From his pulpit at Berlin's university, he molded the intellectual landscape of an entire generation of young Germans, many of whom went on to powerful careers in government and teaching. Treitschke gave intellectual respectability to anti-Semitic agitation that Germany's elites generally dismissed as the unimportant ravings of lower-class mobs.

Indignant criticism and Treitschke's unrepentant responses filled the newpapers for months afterward. A year later, Mommsen entered the fray with his essay.

He turned the language of nationalism against Treitschke and the anti-Semitic mob, accusing them of ripping apart a nation that had only just been put together, of turning every German against his neighbor. As Mommsen put it: Who belonged to "us," and who was part of a foreign element in Germany? Was the only true German a Prussian? A Schwabian? A farmer? In truth they were a motley crew, the German tribes, Mommsen wrote, "and what's it supposed to mean when Herr von Treitschke demands that our Israelite fellow citizens 'become Germans'? They already are, just as much as he is and I am."

All of them, including the Jews, Mommsen continued, carried the characteristic gifts and weaknesses of their own tribes. If they were going to build a nation together, they had no choice but to tolerate each other.

At this point Mommsen uttered the criticism of Jewish behavior that stuck in the mind of the young Fritz Haber. German unity came with a price, Mommsen wrote. It demanded that all Germans give up those loyalties and affiliations that divided them. Germans from Hesse or Hanover were paying that price, and Jews should as well. "Whether they sell trousers or write books, it is their duty, as long as they can do this without violating their conscience, . . . to move with conviction to destroy the barriers that divide them from their fellow German citizens." That meant conversion to Christianity, which according to Mommsen was more of a cultural than a religious identity. "Christendom" now meant simply "civilization."

These were the words, the aging Fritz Haber told his friend, that called him across the threshold of St. Michael's, into the tall medieval sanctuary, past the life-size bronze relief of a glowering Martin Luther. (The church still shelters this Reformation relic, the original lid that covered the great reformer's grave.) With each cascade of water—*in the name of the Father, the Son, and the Holy Ghost*—Haber hoped to become more fully German, to build a more unified nation. And at the time of this conversation with Rudolf Stern in 1926, Haber still felt that his decision was the right one. Germany remained his fatherland, a loyalty that trumped all others.

There's one more story, a bizarre fantasy that Haber told so often that it ended up in the recorded memories of many friends, although no one at the time seems to have given much thought to it. As the story goes, Haber was on a hike through the mountains one day when he got desperately thirsty. Coming upon a village fountain, he plunged his entire head into the water—and at just that moment, so did a nearby ox. When they came up for air, they'd exchanged heads! And from that point on, Haber usually told the laughing crowd, his academic career finally took off.

No one at the time seems to have suspected any deeper meaning in this story. But more recently, the historian Fritz Stern has suggested that it reads as an allegory of baptism. Stern's personal history is intimately entwined with Fritz Haber's—Stern is Haber's godson, and Fritz Haber was a mentor to his parents. Baptism is also an immersion in water, Stern wrote, and it did aid Haber's climb up the academic ladder. So when Haber told this story, did he privately, perhaps even unconsciously, disparage Germany's dominant culture? "Was the ox, known to be a beast of strength and sullen stupidity, the gentile?"

A nother fateful turn of events, another cryptic scrap of paper: This one bears the letterhead of the Technical University of Karlsruhe. Dated December 16, 1894, it states that Fritz Haber had been hired as an assistant in the university's technical-chemical institute. At the age of twenty-six, Haber finally had a paying job.

Karlsruhe, a calm and prosperous city along the Rhine River in southern Germany, capital of the state of Baden, was the perfect place for Fritz Haber. It was a relatively good place to be a Jew, for one thing. Only 3 percent of Baden's voters supported Germany's leading anti-Semitic political party in the elections of 1893 and 1898, fewer by far than in other regions.

The university's chemistry institute wasn't among Germany's most renowned. But it had practical advantages: solid funding from the local government and intimate relations with the country's largest chemical company, the *Badische Anilin- & Soda-Fabrik*, known as the BASF, located just a short ride down the Rhine in Ludwigshafen. Haber knew how to work with industrialists. Those lonely months spent visiting far-flung chemical factories would serve him well.

Karlsruhe's chemical institute was a young institution, relatively unencumbered by formality or tradition. Haber's supervisors gave him considerable freedom, though little money at first. Haber could range as far as his energy and his mind could reach.

Having struggled for so long to grasp this bottom rung of the academic ladder, Haber's grip was all the more tenacious. Over the following seventeen years, the sturdy yellow-brick structure of Karlsruhe's chemical institute became the scene of a startling metamorphosis. Through unending labor and serendipitous fortune, Fritz Haber transformed himself from scientific castaway to conquering hero.

THREE

Ambition

He went at a thing like a bull at a gate.
—W. H. Patterson (English chemist),
describing Fritz Haber

F RITZ HABER'S DEBUT PERFORMANCE on the scientific stage captured Wilhelm Ostwald's full attention and fixed itself in his memory. Years later, while writing his memoirs, Ostwald remembered the scene clearly: the fifth annual congress of the German Electrochemical Society in 1898, hosted by Ostwald himself in his gleaming new Institute of Physical Chemistry in Leipzig—the institute that Haber had once tried and failed to join.

The days of this congress were filled with scientific presentations, and it was midmorning of the second day before the young man from Karlsruhe strode to the podium. Fritz Haber was twenty-nine years old, merely a *Privatdozent,* a position roughly

Fritz Haber in his Karlsruhe laboratory, around 1903.

equivalent to a newly hired assistant professor. "His name was barely known at the time," wrote Ostwald.

Rather than being intimidated, Haber seemed energized by the spotlight. He galloped through his presentation at a breathtaking pace, and many in the audience struggled to keep up.

The details of Haber's presentation aren't so important. (They concerned the circumstances under which electricity, when passed through a liquid solution, breaks chemical bonds of compounds in that solution.) What Ostwald remembered, and what struck everyone else in the room as well, was the young man's energy and apparent self-confidence.

The assembled scientists applauded enthusiastically as Haber wrapped up his speech. Ostwald, chairing the meeting, was both impressed and amused. "Gentlemen!" he said. "Dr. Haber has showered us with such an abundance of material that we won't be able to discuss all of his many and highly interesting points." So after thanking Haber and reminding the audience that lunch was near, Ostwald moved on to the next speaker.

A few in the room, however, seemed to consider Haber's brash performance a personal affront. The next day, a more senior scientist from Bonn tried to put the young chemist in his place; Haber's conclusions, announced the man from Bonn, were overblown and contradicted by other evidence.

Haber leaped to his own defense, and a lively argument followed. Ostwald took Haber's side. "As chairman, I felt compelled to intervene" against this "vehement and unfounded attack," the elder statesman of physical chemistry wrote in his memoirs. Yet even Ostwald, it seems, found the verve and presumption of the young man from Karlsruhe a bit off-putting. Later that year, Ostwald wrote to an acquaintance that "I completely share your view that Haber still needs a bit of seasoning. . . . He's only been in electrochemistry for the last two years, and for that reason [lacks] a sure grasp of the subject."

The scuffle in Leipzig reveals the emerging phenomenon of Fritz Haber, an impetuous scientific outsider fighting for respect and acceptance from sometimes resentful colleagues. It displays his extraordinary energy, quick wit, and ability to command a stage.

Ostwald was correct; Haber had barely set foot in the field of electrochemistry—or the broader one of physical chemistry—and yet he presumed to challenge the field's established authorities. Nor was it the first time he'd mastered a field overnight. A few years earlier, soon after Haber arrived in Karlsruhe, a teaching slot had opened up in the field of dyes and fibers. Haber, who'd never studied dyes and fibers, seized the opportunity and became the university's expert in the field.

He worked, it seemed, day and night, absorbing the knowledge of others, preparing lectures, and conducting experiments that would lead to publications in leading scientific journals. When asked how he managed to master such a range of specialties so quickly, Haber replied that he "studied every night until 2 a.m. until I got it."

Somehow he found time to keep one foot in the practical world. On two occasions, he took off on extended tours of the factories and laboratories of industry, studying their practices and problems. The first tour focused on Germany's world-renowned synthetic dye industry; during the second, he shifted his attention to electrochemistry.

"He taught himself," wrote Haber's later collaborator, the Englishman J. E. Coates, admiringly. But that's not completely true.

Throughout his life, Haber was an intensely social creature. He learned best by talking and listening, rather than solitary study, and many of the steps of his career were determined by the serendipity of personal friendships. So it was in the case of physical chemistry.

In 1896, a young Austrian chemist named Hans Luggin arrived in Karlsruhe from one of the hotbeds of physical chemistry, the laboratory of Svante Arrhenius in Stockholm. Luggin and Fritz Haber became fast friends, and Luggin became Haber's guide to physical chemistry. Haber readily admitted that up to that point, he'd had difficulty grasping many of the mathematical complexities of the field. Luggin "gave Haber what he needed most, the constant exercise and stimulus of discussion," wrote one colleague.

Haber's collaboration with Luggin was cut short, tragically, in 1899 when Luggin became ill and quickly died. He was only thirty-six years old. Luggin's father entrusted Haber with his son's scientific affairs. Haber finished a series of experiments that Luggin had begun and published the results. In the years that followed his friend's death, when Haber wrote scientific articles, he repeatedly noted his professional debt to Luggin.

On the heels of his attention-grabbing performance at the Electrochemical Society congress in Leipzig, Haber published a new textbook on electrochemistry. A senior professor at Karlsruhe had discouraged Haber from taking on the ambitious project, predicting that he would embarrass himself, but the opposite occurred. Influential colleagues praised it as a novel synthesis of electrochemistry's two distinct sides, the practical and the theoretical. The book established Haber as a force within German chemistry.

That success was crowned, at the end of the year, with a promotion. He became an *außerordentlicher Professor*, a status just one step away from every German scholar's ultimate goal of full professorship. The promotion didn't bring any increase in salary, but it was still, to quote German historian Margit Szöllössi-Janze, "an enormous success for someone who'd arrived at the university just four years earlier with only mediocre qualifications."

Yet Haber barely paused to savor his success. He wanted even more. Karlsruhe's chemical institute had received permission to expand, establishing a new chair in physical chemistry, Haber's new specialty. Haber didn't heed the subtle signs that this professorship was intended for someone older and more experienced; he wanted it himself. And when, in 1900, the position went instead to Max Julius Le Blanc, one of Ostwald's former assistants, Haber was bitter. In a letter to Ostwald, he confessed that he felt completely superfluous at Karlsruhe, and would almost certainly have to move on to a different university. (The letter can't be taken completely at face value. Ostwald supervised the shuffling of physical chemists among German universities, and Haber's letter served as an informal application for any other position in physical chemistry that Ostwald might know about.)

Despite his lengthening list of accomplishments, Haber remained an outsider in his chosen field. He felt keenly his lack of an influential mentor, like Ostwald, who could provide protection and intellectual legitimacy. He also knew that baptism hadn't completely erased his identity as a Jew.

Haber compensated by working even harder and asserting himself even more strongly. And like many outsiders, he developed a thin skin, a special sensitivity to slights. He feuded with other scientists, and when criticized he responded sharply. When Le Blanc, Haber's new rival in Karlsruhe, spoke at faculty seminars, Haber regularly found the weakest point in Le Blanc's argument and publicly laid it bare for all to see.

Some resented his ambition and drive. The head of Karlsruhe's chemical institute was known to advise young students dryly that they should take their questions to Haber: "He knows everything. In fact, he knows even more. He's a know-it-all."

Whether because of overwork, inborn personality, or frustrated ambition, Haber's "nerves" started acting up again. His symptoms, as Haber described them in letters, included physical exhaustion, irritability, insomnia, anxiety, and wild overreaction to perceived criticisms. Nearly every year, beginning in the summer of 1898, Haber retreated to a sanatorium or a hot spring resort, hoping to recover his energy and calm his "nervousness." Judging by the complaints in his letters, the positive effects of these retreats didn't often last long.

Fritz Haber had one extremely valuable link to his field's ruling clique: Richard Abegg, his onetime Berlin classmate, the friend who'd first introduced him to physical chemistry. Abegg had moved from Ostwald's laboratory to the University of Göttingen, where he worked as an assistant to another former Ostwald student, Walther Nernst. In 1899, Abegg took a job teaching at the university in Haber's old hometown of Breslau.

Abegg had his finger on the pulse of the scientific community, at least within his particular field of chemistry. Together with Nernst, he edited Germany's scientific journal of electrochemistry.

He knew all the personalities in this fraternity, and heard all the gossip.

Through Abegg, Haber knew what others were saying about him. Abegg smoothed the path of Haber's scientific reports into the pages of the journal of electrochemistry. And when, in the course of some arcane scientific dispute, Haber fired off an overly hasty or ill-tempered letter to the journal, Abegg usually managed to persuade his volatile friend to withdraw it or tone it down a bit.

In Breslau, Richard Abegg also became Fritz Haber's link to a person who belonged to his past and his future. She was a young woman, the first woman ever to acquire a doctorate from Breslau's university. Richard Abegg was her academic adviser. Her name was Clara Immerwahr.

FOUR

Clara

What Haber never understood was women. One day we went walking and the conversation came around to women. He said, "Women are like lovely butterflies to me. I admire their colors and glitter, but I get no further."

—Max Mayer,
friend of Fritz Haber

THE FATEFUL POSTCARD, its stiff paper yellowing and fraying around the edges, bears silent testimony to human passions long since extinguished. The card was still blank on March 14, 1901, when Fritz Haber laid it on his writing desk, added ink to his pen, and boldly addressed it to two old friends and fellow chemists: *"Prof. Dr. Abegg und Fräulein Dr. Assistent, Breslau."* It's likely, considering what happened in the following weeks, that the card was mainly intended for the "Fräulein Dr. Assistent," Clara Immerwahr.

Haber was in the mood for adventure. Or perhaps he was just

Clara Immerwahr.

lonely. The German Electrochemical Society was planning to meet a few weeks later in Freiburg, not far from Karlsruhe, and Haber wanted Richard Abegg and Clara Immerwahr to join him there. Turning the card over, he filled its blank space with a home-made bit of rhyming doggerel.

The mists retreat and fresh new green
sprouts forth in new leaves softly.

Who could resist the urge to flee
from formulas and letters?
And like the free ascending birds
Over imposing mountains
where—see!—eternal flowers bloom
near our Italian cousins.
But if you stay, the way I do,
You need at least to write me.
There's nothing worthier of praise:
Renounce your conscientious ways!

So tell me, wouldn't you like to go to Freiburg?
In any case, *auf Wiedersehen!*

F. H.

The postcard did reach its intended recipient, thirty-year-old Clara Immerwahr, whom Fritz Haber had met many years earlier. And Immerwahr did travel to Freiburg.

In the visible facts of their life, Fritz and Clara had much in common. Both had grown up within Breslau's Jewish community, though neither had ever been religiously observant. Both were children of prosperous merchants, although Clara, through much of her youth, lived on a country estate about fifteen miles outside the city. Clara was two years younger.

The Jewish community of the Habers and the Immerwahrs, wrote one observer later, was an "intellectual aristocracy": educated, witty, humane, politically liberal but socially insular. It was a cozy world, knit by marriages and overlapping friendships. Fritz once had courted Clara's friend Julie Hamburger; Julie was also the sister of his friend Max Hamburger. Fritz had studied in Zurich under the chemist Georg Lunge, who was in turn the cousin of Philipp Immerwahr, Clara's father.

Two bits of evidence suggest that their acquaintance turned to romance in 1891, when Fritz was ending his university studies in Berlin, about to embark on the visits that his father had arranged to chemical factories in eastern Europe. In April of that year, Clara wrote to Max Hamburger, asking him to send her a photo of Fritz. She needed it, she wrote mysteriously, for "really quite simple reasons" that she couldn't at that moment divulge. The second bit of evidence comes from the pen of Fritz, ten years later. In a letter that he wrote in 1901, soon after the postcard inviting Richard Abegg and Clara Immerwahr to Freiburg, he confessed that he'd spent the last decade "diligently but unsuccessfully" trying to forget about Clara.

When Fritz and Clara first became acquainted, Clara was a young woman of twenty-one, in the midst of grief and change. Her mother had died from cancer the previous year. Together with her father (her siblings had left home already), she'd moved from the family's landed estate to a house directly on Breslau's beautiful central square, facing City Hall.

At that time, though, marriage was out of the question. Fritz had no secure job or any immediate prospects for getting one. So he moved on, pursuing his academic career.

Over the next ten years, Clara Immerwahr accomplished extraordinary things, acquiring an education to which she had no right, under the law and customs of the time. She had never attended *Gymnasium,* the essential stepping-stone toward university, because no such schools in Breslau admitted girls. But she covered the same ground with private tutors, then found a small gap in the barriers that kept her out of Breslau's university. Young women could attend university lectures as guests if they secured permission from each professor beforehand.

Step by determined step, Clara pursued her goal, driven, it seems, by sheer thirst for knowledge. Any sober consideration of her future would have compelled her to admit that a nineteenth-century woman had few real prospects in German science, either in industry or in academia. Yet she pressed on. After attending university lectures for a year, she took and passed the examinations that gave her the standing of a *Gymnasium* graduate, then returned to the university in pursuit of a degree in chemistry.

She also converted to Christianity, and was baptized under the distinctive wedge-shaped steeple of Breslau's *Barbarakirche,* a few steps from the house where Fritz Haber had lived during his school years. Like Fritz, Clara was a crosser of boundaries. Her parents put up little resistance to her conversion. The family rarely, if ever, ventured into the synagogue, and Clara's father was considered a free-thinking humanist.

In 1899, fate brought Fritz Haber's former classmate and good friend Richard Abegg to Breslau. He took a position teaching chemistry at the university, and became Clara Immerwahr's academic adviser. The two developed a friendship that was both properly formal and heartfelt.

It is only through Richard Abegg, in fact, that we begin to understand Clara Immerwahr. For years, until Abegg's untimely death, the two stayed in touch through letters, and those letters shed some light on the secrets of Clara Immerwahr's heart. Only her side of the conversation is preserved; Abegg's letters were lost or destroyed.

Those letters were Clara Immerwahr's lifeline during long months when she carried out research at the laboratory of a mining academy several hours from Breslau. From afar, Abegg tried to provide encouragement and advice regarding her experiments. The experiments were complicated and tedious; Immerwahr was measuring the extent to which certain salts—of copper, lead, cadmium, zinc, and mercury—dissolve in liquid solution, forming ions. A lone woman surrounded by men, none of them great supporters of women in science, Clara sometimes was overcome by self-doubt and despair. And Abegg's admonishment that she have "fresh courage and nerve" simply made her feel worse.

"I know that Herr Professor meant well," she wrote, using the formal style of addressing a superior, avoiding the word "you." "But in this case as so often, it confirms the old saying: The well-fed can't understand the hungry. And one doesn't make a sad person happier simply by telling him, 'Cheer up!' . . . I've only known the joy of life in fleeting moments, and I may say that each of them has been balanced out by years of heaviness. How am I supposed to have "fresh courage and nerve" when it takes all my meager strength just to cope with daily existence? . . . I write this to Herr Professor not as an accusation, but because I cannot bear to keep carrying bitter feelings in my heart toward people who are dear to me."

The emotional storm passed. Abegg remained Clara's irreplaceable mentor and friend. When a rumor reached her ears that he might take a job elsewhere, she flew into a panic at the prospect of being "orphaned." And after Abegg assured her that the rumors were false, she was overjoyed. "I don't need to explain to Herr Professor what this means to me, after all that's been said before. In any case, I'll have to perform some good deed today, and in the accounts that I keep with Fate, I've crossed out the whole last page of debits and added a big entry on the credit side." Then, after four pages of detailed information about her experiments, her emotions spilled into a giddy series of teasing postscripts.

p.s.$_2$ Why does Herr Professor have no address stamp, anyway?

p.s.$_3$ Why exactly does Herr Professor degrade me to "gracious Fräulein"?

p.s.$_4$ If I'm particularly impudent today, it's only because I'm so terribly happy!

Three days before Christmas in 1900, a crowd composed mostly of women crammed into the ornate ceremonial auditorium of Breslau's university, the Aula Leopoldina, to watch Clara Immerwahr defend her dissertation. "Seldom has the awarding of a doctorate been attended by so many," noted a report in the Breslau evening newspaper under the headline "Our First Feminine Doctor."

Two students, young men, played the role of Clara's opponents at this academic trial, posing questions about her research and probing for weak spots in her knowledge. According to the newspaper's anonymous correspondent, they didn't allow the "duties of chivalry" to hinder their vigorous questioning. The young woman at the center of the day's festivities, however, "a dainty figure with short blond hair and a spirited expression," proved both skillful and quick-witted in defense of her research. Finally, the dean of the philosophical faculty presented Clara Immerwahr with her diploma, the first one that the university had ever awarded to a woman. "Science welcomes each person, irrespective of sex, confession, race or nationality," he announced, and proclaimed his joy to see this *doctissima virgo*, this "most learned young woman." Still, he "does not wish to see the dawn of a new era"; women, he said, should continue to find their most beautiful and holy duty within the shelter of the family.

Dr. Clara Immerwahr stayed at the university and became Richard Abegg's laboratory assistant. And this is where, a few months later, Fritz Haber's postcard found her. Two weeks later, another letter from Haber arrived, repeating his desire to see Abegg and Immerwahr during the upcoming conference in Freiburg.

Clara Immerwahr joined Fritz Haber in Freiburg. And there, over the course of just a few days, Haber persuaded her to link her life with his.

Fritz was insistent; Clara, filled with misgivings. According to stories that Fritz told later, Clara initially turned down Fritz's proposal, saying that she "wasn't the right sort for marriage." And

Fritz's own letters contain hints of Clara's reluctance. "Fate has been good to me," wrote Fritz Haber to an uncle of his betrothed, announcing his engagement. "Your niece . . . has accepted my proposal. We met at the congress in Freiburg, we talked, and in the end Clara was prevailed upon to give it a try with me."

Clara, meanwhile, in a letter she wrote eight years later, portrayed herself embracing marriage cautiously, almost as a form of scientific exploration. "It was always my approach to life that it's only worth living if one develops one's abilities to their fullest and experiences all that human life has to offer, to the extent that one can. And so I decided to get married—among other reasons—because I felt that otherwise a page of the book of my life, and a chord in my soul, would lie fallow and untouched."

The newly engaged pair traveled immediately to Breslau "like a fairy-tale prince and princess, caught up in a dream," in Fritz's words, to announce their intentions and acquire the traditional assent and blessing from Clara's father. Sudden tragedy, however, overshadowed their good news. Clara's brother-in-law, after picking them up from the train station in a coach, was stricken on the way home by a heart attack and died. A few months later, in August of 1901, Fritz and Clara were married.

For Fritz, marriage was one more step along the path he'd already been traveling toward stability and social stature. It represented no fundamental shift in direction. He threw himself into research with even greater passion. For Clara, on the other hand, it provoked a crisis of identity.

In leaving Breslau behind, and moving to Karlsruhe, she also abandoned her tenuous position in the world of science. No longer a pathbreaking young scientist, she became a professor's wife, responsible for running a household, cooking, cleaning, washing, and mending. She could visit her husband's laboratory and read the latest scientific journals, but she was only an observer in that world, not a participant.

Some people have a gift for carrying their cares lightly; worries scatter from their minds like dry snow off a speeding car. Clara Haber was not so fortunate. Fears stuck to her, and weighed her down. Nor was she blessed with the carefree combativeness that might have allowed her to challenge the expectations of society— or her husband. Fritz may have found the idea of an intelligent and educated wife entrancing, but ultimately he wanted a traditional household even more. So Clara, sensitive and brooding by nature, was trapped, unable to pursue her intellectual passions and unable to find satisfaction in her newly assigned role.

To her mentor Richard Abegg she wrote of "drowning" in sewing and other household work. She asked plaintively why she hadn't heard from him or his wife for so long. The dream of returning to science still flickered in her mind, but its realization would have to wait "until we're millionaires and surrounded by servants. Because I can't completely leave it behind, not even in my thoughts."

Then, a few months later: "I'm working every afternoon in the institute now, reading and preparing technical drawings. I'm doing better now. For quite a long time I was feeling down, but I think it was a physical thing."

There were, to be sure, happy times. One amusing letter to Abegg takes the form of a dictation delivered by Fritz and recorded by Clara, who added her own playfully subversive remarks concerning the "dictator." And Fritz was always lively and high-spirited when he invited students or colleagues from the institute to their home, as he liked to do. Yet Clara Haber's letters to

Abegg display a striking contrast in tone. When discussing chemistry, or professional disputes among colleagues, her writing is animated, lively, and confident. When the topic turns to domestic and personal affairs, she seems frustrated and frequently seized by dark moods.

As 1902 dawned, Clara was pregnant and worried, tortured by the memory of a colleague's wife who had died in childbirth. "I told Fritz recently, I'd rather write ten dissertations than suffer this way," she wrote to Abegg.

The baby, a son, arrived on the first of June. His parents named him Hermann, and had him baptized. "My wife is doing satisfactorily; the frog is fine," Fritz wrote to Abegg. The baby was doing better, at least, than his father. The accumulated stresses of professional feuds and worries about both mother and child had brought on a bad case of colitis, leaving Fritz in agony. Clara had two patients on her hands, not counting herself.

For the next half year, at least, Fritz Haber wasn't going to be of much help when it came to raising Hermann, and not just because of physical ailments. A great honor had dropped into Fritz's lap, an assignment that he certainly wasn't going to turn down. He'd been asked to tour the United States as the official representative of the German Electrochemical Society. It was to be a four-month-long information-gathering trip, sizing up that youthful giant among nations.

On August 18, shortly after his and Clara's first anniversary, Fritz Haber sailed for America. Clara, left to cope on her own with ten-week-old Hermann, retreated to her father's house in Breslau.

The demands of Haber's profession pulled him away from his family, but he also, of his own accord, fled. From this beginning and throughout his life, Fritz Haber never really found domestic peace or a stable balance between professional and family life. In comments that masqueraded as jokes, he sometimes spoke of family as something confining, as the enemy of true friendship and the "murderer of talent." He loved to play the expansive host, welcoming guests to his house and dinner table, but when it came to everyday responsibilities, he often escaped—into his work, into travels, and into the sanatoriums where he hoped to calm his agitated nerves.

FIVE

The Enthusiast

Singing my days
Singing the great achievements of the present,
Singing the strong light works of engineers,
Our modern wonders, (the antique ponderous Seven outvied,)
In the Old World the east the Suez Canal,
The New by its mighty railroad spann'd
The seas inlaid with eloquent wires.

—Walt Whitman

JANUARY 18, 1903. BERLIN

M EINE HERREN!"

In the lecture hall of Berlin's Hofmann-Haus, where Germany's chemical industry gathered, all eyes turned toward Fritz Haber. The young chemist, just thirty-five years old and already nearly bald, stood ready to report on his odyssey through a distant and possibly threatening land.

Fritz Haber with his assistants and laboratory workers outside the Karlsruhe university's Institute of Chemistry. Haber is the bald man seated in the center of the front row.

He'd played two roles, of ambassador and spy, on this journey to the United States. Officially, he'd been an emissary from Germany's chemical establishment, responsible for improving friendly ties that spanned the Atlantic. Unofficially, he'd been sent to take the measure of a rival.

These were the days of misunderstood Darwinism. People heard the phrase "survival of the fittest" and imagined a similar struggle for survival among nations and civilizations, as if each German or American individual represented one cell of a greater organism that was fighting to reproduce and survive. In Germany, the idea mutated into another popular phrase: "the great struggle among nations."

"For a long time, we underestimated the advances of the United States," Haber told his audience. "Now we've gone to the other extreme, and our unshakeable confidence in German superiority has been replaced by a common and frequently unjustified fear that we'll be overtaken in all areas."

In another speech a few months later, Haber put it even more succinctly: "The American threat has become a slogan, and it seems that Bismarck's phrase—that the Germans fear nothing but God—is being gradually amended in business circles with the additional phrase: And the United States, a little bit."

In reality, though, Haber didn't regard the people he'd met in the United States as rivals. He'd met soul mates, fellow adherents of a modern faith in technical progress.

Gentlemen!" he called again. "Anyone who arrives for the first time under clear skies in the harbor of New York is struck by a marvelous impression. Landscapes may be more charming elsewhere, but I have never witnessed such a bustling scene of economic activity. That impression grows as one enters the inner harbor and becomes overwhelming in the lower city, where the means of transportation dwarf anything with which we are acquainted in the old continent."

The Brooklyn Bridge, already twenty years old, reigned over the East River. On the city's grandest avenues, elevated trains rumbled overhead while trolley cars clattered along underneath. Urban life still ran on rails; the automobile was yet to come. In the countryside, horses remained the engines of agriculture and commerce. Beneath Haber's feet, workers already had hollowed out snakelike tunnels for the next miracle of movement, the subway. A first generation of skyscrapers, the Flatiron Building on Broadway and the Park Row Building on City Hall Park, towered 300 to 400 feet into the air, and thousands of electric lightbulbs turned them into shining beacons at night.

That first impression casts a spell, Haber told his Berlin audience. It overpowers the mind. Under the shock of this sensory assault, a European visitor is prone to imagine equally stupendous American factories, universities, and intellects, superior to anything European. With difficulty one remembers certain facts: America does not, after all, rule the world. It cannot be, from one end to the other, so awe-inspiring.

With that, Haber launched into an account of his travels across America. For sixteen weeks he'd been under way, covering thousands of miles from Atlantic to Pacific and back again. He'd visited the universities of Boston and Philadelphia, the aluminum and carbide factories of Niagara Falls and Sault Ste. Marie, copper and lead refineries in Montana and in British Columbia, and hydroelectric power stations in the Sierra Nevada mountains. After stopping in San Francisco, a picturesque boomtown awaiting imminent obliteration in earthquake and fire, he returned east by way of Arizona, Denver, Chicago, and West Virginia. His route did not touch the largely agricultural states of the vanquished Confederacy.

Despite barely speaking English—fortunately, many American chemists had studied in Germany and could explain what they were doing in his native tongue—Haber turned out to be a remarkably competent industrial spy. He had plenty of practice visiting unfamiliar factories, of course, starting with the industrial apprenticeships that his father arranged upon his graduation from university. But the task also fit Haber's personality: He thrived on conversation, his mind took in and digested new information at a staggering pace, and he found everything fascinating, from the prevalence of aluminum-based paint on American mailboxes to the widespread fears of impure water.

Back in Berlin, his report overflowed with technical details about the factories he visited, including, as it turned out, company secrets. Some American executives thought that Haber had agreed to keep their conversations confidential. When an expanded version of his report was published, they were dismayed to find their operations laid out precisely for all the world to see. Word of this got around; more than thirty years later, a British chemist still remembered Haber's "persistence in noting out things in America to an extent that was sometimes not quite agreeable." The chemist put it down to Haber's Jewishness, "the Jews being full of ambition to know and get on and miss nothing."

More impressive than Haber's passion for data collection were his quick and accurate assessments of the American economy. Many Germans at the time believed that cheap hydroelectric power from America's mighty rivers had powered the nation's industrial rise. Haber carried out some rough calculations and concluded that this could not be correct. The key, he decided, was the country's abundance of cheap coal—cheap especially in comparison with the high wages paid to workers, who seemed in chronically short supply. The result was a constant drive to save on labor costs and use more machines. A century later, economic historians generally agree that Haber's amateur analysis was correct.

But what impressed Haber most deeply was America's bent toward the practical. Americans, he said, possessed an instinctive feel for machines and technology. "The atmosphere of life is soaked in mechanical-technical ideas; without any conscious effort, the eye of each person is trained, early on, to look at things in a technical way. We Germans, educated in more abstract style, are forced to learn this later in life with great effort." And while German professors concentrated on their theories, every American scientist tried to produce something of practical importance. Industrialists prowled the halls of academia looking for new ways to capture profits.

In this respect, at least, Fritz Haber was already the most American of imperial German scientists. He, too, was an impatient man of action, drawn to projects with immediate practical consequences. He hoped for a synthesis of the two scientific styles— German depth married to American practicality—in pursuit of a dream of progress that united both nations.

One of the few who recognized the similarities between Germany and the United States a century ago was Adolf von Harnack, adviser to Wilhelm II on matters of science. "Geographically," he said in 1907, "America is for us among civilized countries the most distant; intellectually and spiritually, however, the closest and most like us."

Both were upstart nations, fighting for respect, yet convinced that they owned the future. Each, in its own way, fought to equal and surpass the more established empires of England and France. And the main tool of that pursuit, for both nations, was technical innovation. Progress—or on the other side of the Atlantic, *Fortschritt*—was a secular faith that united Democrats and Republicans in the United States, nationalists and socialists in Germany.

No nations on earth took to the machine with greater enthusiasm. They adopted railways and telephones more rapidly than any of their neighbors. Fascination with technology shaped national cultures and self-perception. In the United States, the belief in Yankee ingenuity took hold. One American novelist, awestruck at the machines on display at Philadelphia's Centennial Exhibition in 1876, wrote that "it is still in these things of iron and steel that the national genius most freely speaks; . . . the present America is voluble in the strong metals and their infinite uses."

In Germany, a new style of behavior emerged, one uniquely attuned to the needs of the machine. Strange as it may seem today, Germans once were noted for their rather sleepy pace of life. As one army general complained to the national parliament in 1891, clocks in schools often were set back ten minutes "so that the students are there when the teacher arrives." Managers at steel factories of the renowned Krupp firm felt it necessary to remind workers that it was against the rules to take naps during working hours. A similar machine-driven transition happened in other countries as well, but it happened more quickly in Germany, and with greater enthusiasm. By 1900, Germans had shifted gears; for most of the next century they would set the

Western world's standard for punctuality, precision, and human organization.

Germans, more often than Americans, subscribed to a kind of technological militarism. They saw innovation as a weapon, a tool of national survival and supremacy. Carl Engler, rector of the Technical University in Karlsruhe and Fritz Haber's most powerful mentor, put it this way in a speech in 1899: "No nation can withdraw from economic competition, the pursuit of technology and advancement of industry, without putting its very existence at risk. . . . [T]he struggle for existence—the fate of the nation—is decided not just on bloody battlefields, but also in the field of industrial production and economic expansion."

Americans, meanwhile, seemed to pursue technical innovation more often for the sheer thrill of it. The thrill was heightened, in many cases, by the seductive allure of quick riches.

In both countries, devotion to technical progress took on a life of its own, and grew into a cause to be pursued for its own sake. Train riders came to expect ever-faster travel; speed represented progress. Railway engineers, who best understood the limitations of their machines, learned to despise the cult of speed. The public, they felt, seemed caught in the grip of a collective psychosis.

Thomas Edison and Alexander Graham Bell became celebrities and secular prophets. In Germany, privileged young aristocrats took to dreaming of new machines "the way earlier generations dreamed of the hunt."

The tide of enthusiasm for technical innovation began to overwhelm long-standing levees of financial caution and tradition. Enormous projects such as central electrical works and canals came into being as though carried forward by irresistible political forces, transforming lives and landscapes along the way. A century earlier, doctors had taken decades to accept the relatively trivial innovation of the stethoscope; around 1900, X-ray machines swept rapidly into medical practice and were employed with breathtaking recklessness, as though X rays were as harmless to a patient as ordinary light.

The furious pace of technological innovation also provoked fears and apocalyptic visions of disaster, to be sure. Yet few honestly felt that progress could be stopped, or even slowed. And technology became part of everyone's dream for a better world, even when they disagreed about what would make the world better. "The nationalists needed technical progress to strengthen the empire, the socialists in order to create the future state," writes one contemporary Germany historian.

Along the Spree and the Rhine, the Potomac and the Mississippi, progress offered people hope for the future. It was a new and optimistic faith, and Fritz Haber professed it with full conviction.

B ack in Karlsruhe, Haber resumed his frenetic pace of research. "Fritz is so scattered, if I didn't bring him to his son every once in a while, he wouldn't even know that he was a father," wrote Clara.

The breadth of his interests continued to amaze and confound his colleagues. During 1904 and 1905, Haber published seventeen different papers in half a dozen different journals. Some of his investigations were of interest only to his fellow scientists— explaining why electrical currents sometimes caused clouds of metal dust to appear in liquid solutions, for instance, or his determination of the exact cascade of chemical reactions taking place within the flame of a Bunsen burner. Others were intensely practical. He devoted great effort to figuring out how the new tramways of Karlsruhe and Strasbourg released wandering electrical currents in the earth, and how this caused water and gas mains to rust more rapidly. This project, wrote one collaborator, was "a good example of the intense interest and great satisfaction which Haber always found in working for the common good." In collaboration with the company Zeiss, Haber developed new laboratory devices for the detection of different gases, and negotiated a deal that gave himself a portion of the proceeds from any commercial sales.

He wrote one more book, his last and most successful one, called *Thermodynamics of Technical Gas Reactions*. It was a quintessential Haber production, clearly and elegantly composed in the form of lectures, as if Haber was speaking directly to his readers. In the book, Haber tried to convince a skeptical community of chemists that they really could profit from the difficult and highly mathematical theories of physics, in particular from the second law of thermodynamics. The first law of thermodynamics states that energy cannot be created or destroyed; it can only be shifted from one form to another—from the chemical energy stored in wood, for instance, into the heat and light of fire. According to the second law, however, *useful* energy inevitably gets lost in the course of any chemical reaction or physical "work." Haber took this law and used it to explain what actually happened in several important chemical reactions.

In one respect, though, Haber eventually came to see the book as a great disappointment. A year after its publication, the chemist Walther Nernst published his "heat theorem," which came to be known as the third law of thermodynamics. Much less famous than the first two, this law states that there is a temperature, called absolute zero, where the loss of energy due to entropy ceases. Absolute zero cannot be reached in practice, but it became a kind of reference point for chemists, like a distant star that helps one navigate across an unfamiliar landscape, and Nernst's insight eventually won him the Nobel Prize for chemistry.

To the end of his life, Haber felt that he, and not Nernst, might have become the discoverer of the third law. While writing his book, he'd come close to it. One more intellectual leap, and he might have formulated it himself, and achieved scientific immortality. Haber never quite forgave himself for the missed opportunity. His jealousy added fuel to a rivalry with Nernst that became increasingly bitter in later years.

H aber still wasn't a full *(ordentlicher)* professor, and it gnawed at him. Not that he was falling behind others of his age; many others—his friend Richard Abegg, for instance—waited longer to reach the top rung of Germany's scientific ladder. Yet Haber had become by this time one of the most accomplished physical chemists in all of Germany. He was publishing research more often than most full professors. And still he waited in vain for the call to occupy a professorial chair. In the silence, he smelled the scent of prejudice and small-mindedness.

"In my experience, it's been very difficult for me to get a chair anywhere," he wrote to a colleague in June of 1905. "Religion and—if it's not immodest to say this—achievements both stand in my way." In many of the best positions, he wrote, Jews weren't welcome; in other universities, no one wanted to disrupt the local pecking order by recruiting a new professor whose accomplishments far outshone those of the existing faculty.

He really didn't need much, Haber continued, sounding mournful: just a laboratory, a few students, money for equipment, and a respectable salary. "And last but not least I have to be independent. Because I don't want to be anyone's assistant." Judging by this and other letters, it wasn't so much money that Haber longed for; it was the status of a professor's title and control of his own destiny.

Haber's prospects took a decisive turn for the better in 1906. Max Julius Le Blanc, the professor of physical chemistry whose appointment in Karlsruhe had thrown Haber into depression six years earlier—and whom Haber tormented in faculty seminars—left to take a position in Leipzig. The way seemed clear for Haber to inherit Le Blanc's position. Indeed, Haber believed that he'd been promised the job if Le Blanc ever left.

A tortuous selection process rolled into motion. A commission was assembled; candidates ranked. Two candidates emerged at the top of the list: Haber and Fritz Foerster, who was already a full professor at the university in Dresden.

Haber had made some enemies in Karlsruhe, perhaps through his know-it-all style and obvious ambition. This came back to haunt him. The faculty commission declared that the scientific qualifications of Haber and Foerster were equal in merit, but voted three to two in favor of Foerster "in view of his personal qualities." An expression of anti-Semitism? No one will ever know. The university senate immediately accepted the commission's recommendation.

Yet Haber also had powerful friends, none more well-connected than former university rector Carl Engler, director of the chemistry institute and former member of the parliament in the German state of Baden. Engler appreciated Haber's drive and talent. He also knew how to "spin threads to all sides and cleverly tie them fast." Barely a week after the commission's vote, when the matter moved on to the education ministry of Baden, Haber's name suddenly reappeared at the top of the list of candidates. The unexplained reversal carried no fingerprints whatsoever; it was probably the result of Engler's quiet intervention. The ministry immediately offered Haber the position, and Haber, just as quickly, accepted. On August 10, 1906, Grand Duke Friedrich, monarch of Baden, signed the official order appointing Dr. Fritz Haber to the lifelong position of professor in the civil service of the German state.

At the age of thirty-seven, Haber slipped into his new role easily. He savored every taste of new status and authority. It was the life he'd long imagined, and it fit him well.

For the first time since leaving his parents' home, Haber had money to spare. His salary jumped by 50 percent upon his appointment, then increased by another 30 percent a year later, when Haber turned down an offer from the University of Zurich, in Switzerland. Finally, he could genuinely afford the spacious dwelling that he, Clara, and Hermann occupied on the northwestern side of Karlsruhe. They were soon joined by a dog whose "true nature precluded obedience," who led Haber on many a frustrated chase through the city streets.

His scientific kingdom covered half of one floor in the substantial yellow brick building that housed the university's chemistry laboratories. It stood proudly—and still stands—at one corner of a courtyard that forms the historic heart of Karlsruhe's Technical University.

A photograph from 1909 shows Haber with his entourage of thirty assistants and students arranged alongside the building. Haber, almost completely bald, sits in the position of honor in the middle of the front row. He leans back slightly, in calm repose, his hands clasped in his lap.

Some professors found teaching a distraction, and the demands of students a burden. With Haber, it was just the opposite. Students energized him. He worked best in front of an audience, and there was no better audience than a cadre of dutiful and respectful young scientists. He worked out his own thoughts through conversation, sometimes in the course of long walks through Karlsruhe. Haber also relished the role of scientific patron, shaping young minds and careers.

"He didn't restrict his advice to working hours in the laboratory," recalled one student, years later. "Sometimes he'd grab me on the street, and like a spider trapping a fly in its net he'd entangle me in the sticky, inescapable threads of his far-reaching suggestions." Young scientists, usually anxious to bring their dissertations to a successful conclusion, came to fear any consultation with Haber. Instead of answers, Haber showered them with questions and suggestions.

When a problem in the laboratory required his attention, Haber "first of all paced back and forth restlessly, thinking hard, his torso leaning forward, his hands on his back, a pitch-black cigar between his teeth. As soon as he started to talk, he laid the cigar to one side, one end glowing and the other end bitten into an odd brush-point. The cigar always remained behind when he finished and disappeared in a rush—he always moved in a kind of trot." Haber's students began to collect these half-consumed cigars, each one sealed in a test tube. At a festive meeting of the

Karlsruhe Chemical Society, they unveiled this tribute to their leader: a shining wreath of glass-clad cigar remnants, framing a selection of their most successfully drawn caricatures of Haber. Haber led the applause.

A natural extravagance blossomed. Haber seemed to recognize no limits either to his time or his talents. He was ready to explore every scientific question, share money with anyone who needed it, drink and talk with his students at a nearby beer hall—a sign above their usual table read "Lying is allowed here"—with no need ever to watch the clock.

"Haber loved to go out with his people," recalled J. E. Coates, an English scientist who worked in Haber's laboratory for several years. "Sometimes he was like an actor. He used to tell stories, always very long and with many details and decorations. He gave the impression that he was a great man, but friendly at the same time."

One student described him as always "high-spirited and amusing." Another, an American named Gorton Fonda, was surprised to find Haber "so simple in manner, informal and genial. . . . He was always thus among his students, kind, joyful and exuberant, although his home life was not too happy."

Home life, of course, meant Clara. Fonda knew her only from dinner parties at the Haber home, a quiet figure in the background tending to her young son Hermann while Fritz entertained the guests with an endless series of stories, jokes, and rhymes composed on the spot. But Fonda knew her reputation: "It was said that Haber was under continuous nagging from his wife and that he was disturbed by it."

It was obvious to many, apparently, that discord and disappointment filled the marriage of Fritz Haber and Clara Immerwahr. And most of their acquaintances—or at least those whose
memories ended up in archives—blamed Clara.

How could they feel otherwise? They knew Fritz Haber well,
and he seemed a perfectly attentive and gracious husband. Yes, he
might be lost in his work at times and gone a great deal, but that
was the normal life of a university professor. Fritz was fulfilling his
proper social role as they understood it; why could Clara not perform hers with equal good cheer? It seemed clear to them that the
problem lay within Clara, in her anxieties, her melancholy, and
her refusal to play the social role that was expected of a professor's
wife. As one acquaintance wrote: "He suffered a great deal from
her pettiness."

One contrary note survives, from the hand of Paul Krassa, one
of Haber's students. He also happened to be Clara's second
cousin, and the two became good friends when Krassa arrived in
Karlsruhe. "She completely recognized the outstanding talents
and personality of her husband, but it certainly was not easy for
her to be the wife of a 'great man,'" Krassa wrote later. "She sacrificed her profession for him, and she never really found the necessary substitute for it in family life. She had no interest in playing
a prominent social role, nor was she particularly good at it."

At an important university ceremony, Fritz Haber was scheduled to give the keynote speech. "I assumed that Clara would be
up front," recalled another scientist's wife. "But she was hidden all
the way in the back. He would have liked her to be more socially
prominent." This same woman recalled her astonishment upon
finding Clara one morning in the kitchen sitting with her domes-

tic servants, drinking coffee from a broken cup. "Perhaps Clara Haber lacked a certain dignity, but I remember that she really was good to her people."

Clara saw limits and dangers where Fritz, at least during his exuberant phases, saw none. His inability to economize, either with time or money, produced constant friction. Clara worried when Fritz worked himself to the point of exhaustion; she grew angry when he lost track of time and showed up for dinner hours late; she disapproved of Fritz's free-spending ways.

As Fritz Haber and a number of friends returned from an excursion in a steamship during which everyone had gotten soaked by rain, someone asked where Clara was. Haber replied, "She's at home thinking about which family member might have caught what illness in which way!"

Acquaintances found Clara emotionally rigid and prone toward absolute moral judgments. "I felt that she was a person with acute ethical standards, fanatically held," wrote one of Haber's students. A friend of Haber's family from Breslau hinted at the same trait: "She was fanatical about the truth. Nothing could remain unspoken" or ambiguous.

Clara's simplicity of dress, and her quiet rejection of social convention, carried traces of Germany's counterculture, the "life reform movement" that reached the peak of its influence in the years around 1900. This movement had many manifestations, from vegetarian diets and herbal medicine to a rejection of social hierarchies, Prussian patriotism, and corsets.

From Clara herself, few words are preserved from these years, as her husband climbed to prominence. The letters to Richard Abegg grew less frequent as the years passed. In 1907 there was

just one, a postcard in which Clara complained bitterly that she'd
been unable to join Abegg at a scientific conference in England because she'd come down with diphtheria. "And it would have been
so interesting to be in England just now! And I would have so enjoyed having a real conversation with you again."

One of Fritz Haber's friends, while congratulating him on his
appointment to a professor's chair, suggested that he take it
easy for a bit; he'd earned the right to rest on his laurels. Haber
thanked him for the advice, but confessed that he probably
couldn't follow it: He'd become like a machine that, once kicked
into action, always ran at its accustomed pace.

That pace wore Haber out. Again and again, his body or his
nerves rebelled. Intestinal problems drove him to doctors and exhaustion delivered him to sanatoriums. But it was his accustomed
pace, and he couldn't alter it.

Fritz Haber might have continued at that pace for a lifetime,
churning out research, filling the pages of scientific journals, ce-
menting his reputation as one of Germany's most productive
minds. He might even have been satisfied with that life; it fulfilled
his boyhood dream.

But another turn in his fortunes lay just ahead. Coincidence—
a chance contact, a fortunate business relationship, and a scientific
feud—already had set him on the path toward a discovery that
would shape an epoch.

SIX

Fixation

Haber will go down in history as the ingenious inventor of the process for combining nitrogen with hydrogen . . . as the man who by this means won bread from air and achieved a triumph in the service of his nation and all of humanity.

—Max von Laue,

in an obituary published soon after Haber's death

JUST INSIDE the door of San Francisco's Maritime Museum stands an exquisite wooden model of the largest sailing ship ever built, the *Preussen* (in English, *Prussia*). The steel-hulled *Preussen* was longer than a football field, measuring 407 feet from end to end. Thirty billowing sails hung from five towering masts, catching any breath of wind, pulling her smoothly through the waves. A German shipbuilder, so the story goes, constructed the *Preussen* in 1902 to satisfy the whim of Germany's last emperor, the erratic Kaiser Wilhelm II, who loved ships and insisted that Germany have the biggest, fastest, and grandest vessels afloat.

Haber–Le Rossignol apparatus for the synthesis of ammonia, built in 1909, now on display in the Deutsches Museum, Munich.

For eight short years at the dawn of the twentieth century, from 1902 to 1910, the *Preussen* reigned as queen of a commercial armada that navigated the toughest trade route in the world. Fourteen times she departed from Hamburg and set sail for the opposite corner of the Atlantic, the perilous tip of South America. Thirteen times, she reached her destination, fighting her way through the headwinds off Cape Horn, crossing into the Pacific, then sailing north again toward the Chilean port cities of Iquique, Tocopilla, and Taltal.

These remote ports bristled with the masts of sailing ships from all over the world, but mainly from Europe. The ships were drawn by one thing only: thick layers of mineral deposits that lay fifty miles inland, hidden within bone-dry plateaus a few thousand feet above sea level. The rock, called *caliche*, held concentrated nitrogen, a substance essential for the waging of warfare and the growing of food.

Miners blasted open the veins of *caliche* and carried loads of the rock to refining plants. There it was crushed and soaked in water to extract the nitrogen-containing mineral, sodium nitrate. When the water in turn evaporated, it left behind a layer of brown, nitrogen-rich powder. Poured into two-hundred-pound sacks, it was ready for the long voyage northward.

The *Preussen* carried nearly eight thousand tons of nitrate on each of her trips home to Germany, one wave in a tide of nitrogen that flowed from Chile to Europe and the United States. Chile's nitrate mines, which began operating on a large scale around 1870, quickly eclipsed an earlier—though more famous, and smellier— source of nitrogen, guano left by birds on Pacific islands.

From 1895 on, Chile's ports delivered more than a million tons of nitrate each year. It was Europe's lifeblood and the means of its bloodletting. Farmers spread that nitrate on their fields, where it generated bountiful harvests, while industrialists converted it into gunpowder and explosives for the military.

When the *Preussen* sailed toward Chile for the first time in 1902, Fritz Haber, still relatively unknown, had barely considered the nature of nitrogen. By the time the great ship met its end in 1910, rammed by an errant steamer in the English Channel, the nitrogen trade was doomed, for Fritz Haber had created in his laboratory the promise of nitrogen stockpiles vaster than all the mines of Chile.

N itrogen, unseen foundation of life, has two natures. One is ubiquitous, inert, and suffocating. The other is scarce, explosive, and life giving.

The founders of chemistry identified the first form of nitrogen near the end of the eighteenth century when they tried to figure out what chemical substances made up air. Most of air, they realized, simply took up space. Unlike oxygen, this substance did not burn. Unlike carbon dioxide, it refused to react chemically with anything else. It remained aloof, an enigma to science. The Germans called it *Stickstoff,* something that suffocates.

In the course of the century that followed, scientists uncovered nitrogen's second face—the life-giving and even explosive one. They discovered nitrogen within gunpowder, and before long they learned to use chemical compounds containing nitrogen in dynamite and TNT. Nitrogen turned up within the tissues of every living plant and animal, and chemists soon realized that no stalk of wheat or human child could survive and grow without it. (Today we know why; nitrogen atoms are involved in photosynthesis, they form essential links in the chain of DNA, and they make up a substantial fraction of all protein.) Justus von Liebig, the pioneering German scientist, even declared in 1840 that the main goal of agriculture was the packaging of nitrogen for human consumption.

Liebig was right about this, but wildly mistaken about where plants get their nitrogen. He thought that the world was awash in this essential nutrient; that plants received plenty of it in the form of nitrogen-containing ammonia that is carried to earth in rainfall. A long-running experiment in the farmlands of England, though, proved him wrong. Crops in these fields benefited more from added nitrogen than from any other fertilizer. (Liebig, an intellectually pugnacious sort, dismissed the results for years as "impudent humbug" produced by a "set of swindlers.")

During the latter half of the nineteenth century, scientists burned away the fog of mystery that surrounded this enigmatic chemical, and explained the reason for its dual nature. An individual nitrogen atom has what you might call three hitching points; like a three-armed creature, a nitrogen atom can form chemical links with three different atoms. It can link arms with three oxygen atoms to form nitrate, or three hydrogen atoms to form ammonia, or NH_3. But two nitrogen atoms can also bind to each other in the tightest of chemical embraces, with all three arms gripping each other in a nearly unbreakable triple bond.

This "diatomic" nitrogen, or N_2, is the form that blankets the earth, making up nearly 80 percent of the air around us. It is of little use, though, to living creatures. The pairs of nitrogen atoms are satisfied with each other; they refuse to be drawn into the playground games of life that require linking arms in turn with hydrogen, oxygen, or other atoms to create amino acids, proteins, or DNA.

Nitrogen as it exists in the atmosphere will fill a balloon, but it cannot feed a blade of grass. It's as nourishing as the seawater surrounding a thirsting sailor. Only the mightiest of natural forces, the blast furnace of lightning, is able to rip apart the contented pairs of atmospheric nitrogen atoms. In a lightning bolt, nitrogen in the air fuses with oxygen, forming nitrogen oxides that fall to earth with raindrops and nourish both forests and grasslands. This reactive or digestible nitrogen is what nourishes life and ends up in military explosives.

Apart from lightning, one other natural path links the two realms of nitrogen—the useless and the essential. It runs through the family of plants that can extract nitrogen from the air surrounding them, nourishing themselves and even adding nitrogen to the soil. These are the legumes—chickpeas, alfalfa, and clover. Their secret lies underground, within irregular nodules that grow on their roots. Bacteria populate these nodules, and through a miracle of nature these microbes manage to carry out in the cool stillness what otherwise demands the brute force of lightning.

For millennia, farmers have recognized this quality of legumes. They've grown crops of chickpeas or clover in order to "strengthen" the soil, making possible better harvests of rice or wheat during the following growing season. For all their accomplishments, however, legumes are modest producers of food. They've never become central pillars of global food production. The major food crops—corn, wheat, rice, potatoes, and even those less famous, such as cassava, bananas, and common fruits and vegetables—have to acquire nitrogen from another source, the digestible nitrogen compounds already present in the soil.

Before the Machine Age, digestible nitrogen could not be created by the human hand; it could only be recycled from the rotting bodies and excretions of other living things, from vegetable compost, animal manure, and human urine. (Lightning, however awe-inspiring, contributed only a small amount.) These stores of nitrogen were limited. Even worse, they continually grew smaller.

In the course of each growing season, crops extracted nitrogen from the soil. When those crops were fed to cattle, three-quarters of the nitrogen ended up in manure. And when that manure returned to fields, most of its precious nitrogen escaped into the air as ammonia gas or washed into nearby streams. The nitrogen in human waste rarely made its way back to the fields at all. As human populations expanded, especially in Europe, they began to live beyond their means, writing check after check against the land's nitrogen account while making wholly insufficient deposits to replenish those stores.

Nitrogen, in fact, was the weakest link in the chain of life, a substance more scarce than water, sunlight, or any other nutrient. Without steady doses of additional nitrogen, either through the growing of legumes or the spreading of fertilizer extracted from mountain plateaus in Chile, fields could—and did—grow infertile. Crops emerged stunted and weak, pitiful as malnourished children, unable to nourish either man or beast.

I n the course of a short September evening in 1898, one eccentric man transformed these bare facts of nature into grim prophecy, a vision of the future that seized the imaginations of millions.

William Crookes, president of the British Association for the Advancement of Science, was an independent-minded scientist who lived from inherited wealth. He'd made his reputation by discovering the chemical element thallium, then set up his own private laboratory in London. In the freedom of his laboratory he investigated any phenomena he pleased, from newly discovered cathode rays to the possibility of communication with the dead.

In his presidential address to the annual meeting of the British Association that September of 1898, Crookes decided to break with custom. He would speak neither about his own research nor about any topic of current public interest. Instead, he told his listeners, he proposed to introduce a new topic, an unfamiliar one "of urgent importance."

With barely a pause, Crookes leaped to his alarmist conclusion, one based, he asserted, on "stubborn facts." England and all civilized nations, he told his listeners, "stand in deadly peril of not having enough to eat." Yet Crookes immediately provided a glimmer of hope, a call to arms that roused a self-important sense of mission among the scientists in his audience: "It is the chemist who must come to the rescue of the threatened communities. It is through the laboratory that starvation may ultimately be turned into plenty."

Having snared his audience's attention, Crookes turned to specifics. The old ways of agriculture would no longer suffice, he announced. For centuries, whenever societies needed more food, they had simply claimed more land, converting grasslands and forests into wheat fields. During the preceding century, this had been carried out on a grand scale. The great North American prairies and vast areas of Russia had been conquered by the horse-drawn plow. And now there was no more land to be claimed. The hungry mouths of humanity, however, continued to multiply. (Actually, Crookes concerned himself only with the potential hunger of "Caucasian bread-eaters," and he helpfully explained who they were: Europeans, North Americans, "the white inhabitants of South Africa," and "the white population of the European colonies.") Each acre of land would now have to produce more. But how?

The answer had a name, Crookes told his audience: nitrogen. Nitrogen fertilizer made it possible to grow more wheat on less land. Yet the planet's stores of nitrogen also were running low. The guano deposits were nearly exhausted. The nitrate mines of Chile were destined to be depleted within a few decades.

One industrial source of nitrogen had come onto the scene by the time of Crookes's speech. Nitrogen-containing compounds could be recovered from the ovens that converted coal into fuel for the blast furnaces of steel mills. Yet this source of nitrogen also couldn't satisfy humanity's growing needs. The looming shortage of nitrogen, Crookes warned his listeners, represented a far more existential danger than the much-feared possibility that England might run short of coal.

Enter the chemist, savior and guide. Chemists knew that inexhaustible reserves of nitrogen filled the atmosphere. They also knew why those reserves remained untapped. They understood the stubborn triple bond that linked pairs of nitrogen atoms together, preventing them from engaging in any other chemical liaisons. It would be a chemist, Crookes told his audience, who one day would find a way to break that triple bond and "fix" nitrogen in more useful chemical forms, such as nitrogen compounds that plants could digest.

It's an odd term, the "fixation" of nitrogen, but it persists today, a century later. Think of it as arranging dates for individual nitrogen atoms, fixing them up with new partners, like hydrogen, whose arms clutch less tightly than another nitrogen atom.

In Crookes's words, the quest to conquer nitrogen's triple bond took on the aura of an imperialist and even racist crusade: "The fixation of nitrogen is a question of the not-far-distant future. Unless we can class it among certainties to come, the great Caucasian race will cease to be foremost in the world, and will be squeezed out of existence by races to whom wheaten bread is not the staff of life."

The Crookes speech caught a monster wave in the surging *Zeitgeist* of the late nineteenth century. It mixed together intoxicating ingredients of its era: imperial arrogance, apocalyptic visions, and technological faith. Across Europe, newspapers covered it, publishers reprinted it, and intellectuals debated it.

The ideas washed through Fritz Haber's laboratory in Karlsruhe. Haber was probably listening in May of 1899 when Carl Engler, rector of Karlsruhe's university and Haber's professional mentor, dedicated the university's new building with a speech that echoed Crookes's fears and hopes. Technological progress is necessary, Engler told his illustrious audience, "for the world needs new means to support humanity's ever-growing millions. . . . One already can see the time coming when the natural powers of the soil no longer suffice to support those millions, and new methods of production will have to replace the old. . . . It is certainly only a matter of time until technology will make possible the artificial preparation of foodstuffs."

Germany's master of physical chemistry, Wilhelm Ostwald, thought for one breathtaking moment that he'd solved the problem. Ostwald wanted to capture nitrogen in the most direct way possible, by persuading it to link arms with hydrogen to form foul-smelling ammonia, or NH_3. Ammonia, the simplest form of fixed nitrogen, could then become the raw material for production of common fertilizers such as ammonium sulfate. The key to success, Ostwald knew, would be finding the right chemical catalyst, something that would help the reaction along. From experience, he knew that heavy metals such as iron and platinum often did the trick.

In 1900, Ostwald mixed nitrogen and hydrogen gas in the presence of iron, and he detected small amounts of ammonia. Elated, he applied for a patent on his discovery and privately informed Germany's leading chemical companies of his breakthrough.

The companies immediately signed contracts with Ostwald that would have made the chemist a very rich man if his triumph had been real. Within months, however, a young researcher from Germany's largest chemical company, the BASF, debunked it. The newly hired engineer, twenty-six-year-old Carl Bosch, demonstrated that Ostwald hadn't created any ammonia; his process merely released ammonia that was already present in small quantities within the iron catalyst itself. Ostwald, red-faced, withdrew his patent application and abandoned ammonia research altogether.

Haber didn't immediately volunteer for this epic quest. He had to be goaded into it with offers of money and insults to his pride.

The money came first. Sometime during 1903 or 1904, Haber got a letter from the Margulies brothers of Vienna, managers and part owners of the Austrian Chemical Works. The industrialists had detected intriguing traces of ammonia in their chemical plant, and wondered if they might have stumbled across a previ-ously unknown—and profitable—way to create this valuable chemical.

It's a mystery why the Margulies brothers turned to Fritz Haber. They may have known his father, Siegfried, who traveled widely in Austria's eastern territories and had many business con-tacts there. If so, one senses in Fritz's initially unhelpful responses the recalcitrance of a son who still regarded his father's guidance as a yoke to be thrown off.

Twice, Haber brushed them off. But the Margulies brothers of-fered to pay him well, and Haber relented.

Haber focused on the scientific question at the heart of the problem, the willingness of hydrogen and nitrogen atoms to com-bine into ammonia, rather than maintain their separate lives. In the language of chemistry, he was trying to find the point of bal-ance, or equilibrium, between the reaction that created ammonia and the opposite one, in which ammonia broke down into free ni-trogen and hydrogen.

If significant amounts of ammonia formed spontaneously when hydrogen and nitrogen encountered each other, it would be an encouraging signal. If there was little, or if most of it converted immediately back into the two separate elements, Haber knew he'd have to give up; he'd be fighting the laws of nature, trying vainly to forge gold from lead like alchemists of bygone centuries.

Locating this point of equilibrium was no easy task. For one thing, all the participants in this chemical dance seemed gripped by inertia. Neither reaction—from a nitrogen/hydrogen mixture into ammonia or in the other direction—happened easily or quickly. The dancers—the atoms of nitrogen and hydrogen—refused to leave their partners.

Like a band leader shifting the music to a faster tempo, Haber added heat, forcing the dancers to spin faster. At 1,000 degrees Celsius, a temperature at which iron glows red, Haber mixed hydrogen and nitrogen inside a platinum pipe and measured the amount of ammonia that emerged. Then he started again with hot ammonia, allowing it to disintegrate into nitrogen and hydrogen, trying to approach the equilibrium point from the opposite direction as well. As Ostwald had done a few years earlier, he placed iron inside the pipe to act as a catalyst—a rogue element on the dance floor, tearing apart comfortable couples and encouraging them to link arms with someone new.

The results delivered a solidly negative verdict. Only under conditions of fierce heat did hydrogen unite with nitrogen, but the union was short-lived; the heat caused ammonia to disintegrate even more quickly. At a temperature of 1,000 degrees Celsius, ammonia accounted for only about one-hundredth of one percent of the mixture. Anyone who tried to extract ammonia from this reaction would be fighting a futile battle against nature.

Haber informed the Margulies brothers of his disappointing findings, published them in a scientific journal, and went on to other things. And that might have been the end of the matter, had it not been for the antagonism between Fritz Haber and Walther Nernst.

It wasn't exactly a rivalry because Nernst didn't consider Haber a rival. Nernst was only four years older than Haber, but he was rapidly displacing Ostwald as the dominant figure in their field. A year before this, he'd formulated his "heat theorem," soon to be called the third law of thermodynamics, which Haber felt had slipped through his own fingers. Nernst had recently been wooed away from Göttingen, one of Germany's finest universities, by an even more generous offer from the university of Berlin. And he was rich; Nernst had sold one of his inventions, a lightbulb, to German industry for a million marks. Nernst felt his superiority and made sure others felt it, too. Haber envied Nernst, and resented him.

In the fall of 1906, a letter from Nernst arrived in Haber's mailbox. Nernst wrote that he'd been struck by Haber's calculation of the ammonia equilibrium point, because it didn't appear to agree with the result that Nernst would have predicted, based on his "heat theorem." Nernst informed Haber that he'd assigned a student to repeat the experiment, but with pressurized gases. (Increasing the pressure increased the amount of ammonia present in the mixture.) These experiments, he wrote, had produced a significantly different point of equilibrium—with less ammonia— than Haber's experiments, and he intended to announce his results at a large meeting of chemists the following spring.

The prospect of public correction by Nernst was intolerable. Haber leaped into action, repeating his experiments in more elaborate fashion, employing different sizes of reaction containers and different materials—iron, nickel, chromium, and manganese—as catalysts to speed the reaction along. His new results still didn't agree with those of Nernst, though they were closer.

As a practical matter, and even as a matter of science, the differences hardly amounted to much. Neither calculation gave anyone much reason to hope for commercial production of ammonia. For Haber, though, and perhaps also for Nernst, the arcane dispute had become a personal duel, a matter of ego and status. In May of 1907, when Nernst presented his results at a meeting of the Bunsen

Society for Applied Physical Chemistry, Haber immediately presented his new and revised calculations.

Nernst, provoked by this challenge, responded by putting Haber in his place. "I would like to suggest that Professor Haber now employ a method that is certain to produce truly precise values," he announced. And after Haber again defended his results as the more accurate ones (as in fact they were), Nernst simply brushed him aside and closed the discussion. "It is unfortunate that so little ammonia is formed in this state of equilibrium, compared with what Haber's highly inaccurate numbers led us to assume," he remarked. "Because otherwise one might actually have considered it feasible to synthesize ammonia directly from hydrogen and nitrogen. Now the situation is much less favorable. . . ."

H*aber's highly inaccurate numbers.* Those dismissive words wounded Haber's vanity and stirred up all his insecurities. They drove him like spurs back to the laboratory, and back to the task of capturing nitrogen from the air. They also provoked another of Haber's characteristic eruptions of "nerves," with accompanying digestive problems, insomnia, and skin rashes.

At that moment, ammonia became Haber's obsession. And simultaneously, other stars of fate swung into fortuitous alignment, pointing the driven young scientist toward his goal.

First, an extraordinary scientist joined Haber's laboratory, a young Englishman named Robert Le Rossignol with a gift for solving practical problems of engineering. Haber didn't know it yet, but the task that lay before him would require ingenious design of new experimental equipment. In this, Le Rossignol proved a master.

Second, Haber acquired an unusual piece of machinery, a compressor that was able to squeeze previously unthinkable amounts of gases such as hydrogen or air into a steel reaction chamber. Haber began to wonder what would happen if he subjected mixtures of hydrogen and nitrogen gas to such pressures, two hundred times greater than normal atmospheric pressure. Nernst had already compressed these gases—but only moderately—in his experiments, and as expected it had produced larger quantities of ammonia. But what if he combined extreme heat—in order to encourage the formation of ammonia—*and* extreme pressure to prevent the ammonia molecules from immediately disintegrating back into hydrogen and nitrogen again? What if he found a really good catalyst, a chemical that could speed up the reaction even further? What if he could create large quantities of ammonia under those conditions, then quickly cool it down so that it turned into liquid form, so it could easily be recovered? The more Haber thought about it, the more he felt it was worth pursuing.

Finally, Haber acquired a new and powerful industrial partner, the Badische Anilin- & Soda-Fabrik (BASF), Germany's largest chemical company, headquartered a short ride down the Rhine in Ludwigshafen. It was an arranged marriage, pulled together with assiduous effort by Carl Engler, Haber's mentor at Karlsruhe's university. Engler, a man of many contacts, was also a member of the BASF's board.

Engler had several reasons for his matchmaking. In part, he played the role of concerned parent, trying to guide an overenthusiastic son toward sensible partners and away from less auspicious liaisons, in this case Haber's alliance with the Margulies brothers of Vienna. Haber apparently considered reviving that relationship, but Engler advised against it. German professors were government-appointed civil servants, he reminded Haber; it wasn't considered proper for them to place their talents at the service of foreign companies.

Engler also suspected that Haber and the BASF needed each other more than either one realized. Haber required industrial-scale funding in order to acquire expensive new laboratory equipment. The BASF, for its part, needed the jolt of Haber's energy and intellect. The company had heard the siren call of nitrogen, and was working rather ineffectually on two different ways to capture this elusive element. One approach used an electrical arc, a kind of artificial lightning bolt, to drive nitrogen into the arms of oxygen, just as naturally occurs during thunderstorms. The other process combined nitrogen with calcium and carbon to form a compound called cyanamide, which in turn released ammonia gas when subjected to steam. Both processes, however, required prodigious quantities of expensive electricity, and neither was destined for practical success—except in places where electricity flowed as freely as the waterfalls that drove hydroelectric turbines, such as Norway.

The courtship between Haber and the BASF was an intermittently rocky affair, as was the ensuing marriage. The BASF wanted Haber to work only on its electric arc method for combining nitrogen with oxygen; Haber demanded that the company also support his efforts to create ammonia. The company eventually agreed, but only "out of personal consideration for my wishes, and not out of confidence in the matter itself," as Haber put it later.

Haber, meanwhile, suffered pangs of guilt for abandoning past partners who'd turned his attention toward nitrogen in the first place. It would be fair and just, he wrote to BASF executives, to share any discoveries with the Margulies brothers of Vienna. Haber asked if the BASF would consider licensing any future ammonia-related technology to the Austrian firm. The BASF summarily dismissed the idea, and Haber didn't raise it again.

At one point in the negotiations, Carl Engler felt compelled to intervene with BASF executives on Haber's behalf, apparently to smooth feathers that Haber had ruffled with his various demands. "I must emphasize that I have no personal interest in the firm's acquisition of Prof. Haber's services," he wrote. "Prof. Haber is a highly industrious fellow, and I expect much further success from his talent and vigor. . . . And since he's conscious of his own worth and—just like the Ostwald school—likes to earn some money, he naturally can't be had on the cheap." If the BASF turned away from Haber, Engler reminded the company's executives, the young chemist's talents might be snapped up quickly by one of its competitors.

On March 6, 1908, Haber and the BASF signed their deal. What's most remarkable about it is its conversion of Haber, ostensibly a German civil servant, into a private entrepreneur. The BASF funded Haber personally, doubling or tripling his already generous salary, and Haber used the money to hire assistants or pay for equipment. The chemical company became Haber's major sponsor, providing his laboratory with as much funding as the university itself. In return, Haber agreed to turn the results of his nitrogen research over to the BASF and to obtain permission from the company before publishing details of his work. In addition, if Haber's work led to commercial production, he was to receive royalty payments equal to 10 percent of the company's net profits from his discoveries.

If Haber needed any further prodding to focus his mind, it arrived in the form of one more meeting with his nemesis, Walther Nernst, two months later. To Haber's astonishment and alarm, Nernst announced that he now believed it might be possible to create ammonia after all; what's more, he was working on a solution to the problem, one that would render the entire nitrate trade obsolete.

As it happened, Nernst's efforts never succeeded. But his bravado threw Haber into a panic. He dashed off a letter to his new paymasters in Ludwigshafen, informing them of the urgent need to pursue ammonia as rapidly as possible.

There followed a pell-mell rush toward triumph. In hindsight, success came with amazing speed. Within a year, Haber and his colleagues overcame two major obstacles, one through ingenious engineering, the other through blind luck.

First, the engineering challenge: They had to figure out a way to keep a mixture of hydrogen and nitrogen circulating through a steel "reaction chamber" while squeezing them under pressures that could rupture steel—up to 200 "atmospheres," or 200 times the normal pressure of the atmosphere at sea level. The compressed mixture also had to be cooled quickly as it left the chamber, and incoming gases heated. Robert Le Rossignol took the lead in solving these problems, and did so masterfully. He and the laboratory mechanic created a new valve that could tolerate such high pressures while still letting gases pass through. The hot gases escaping from the reaction chamber, meanwhile, transferred their heat to the cool incoming gases, accomplishing two goals at once.

There remained only the search for a better catalyst—a material

that could, through mysterious processes still not completely understood, grease the wheels of this particular chemical reaction.
Both Haber and Nernst had used iron as a catalyst in their earlier
attempts, and it worked, but not well enough. There was nothing
to do but play chemistry's lottery, trying out one material after the
other, hoping for a lucky strike.

Once again, Fritz Haber profited from coincidence, in this case
a budding professional relationship with another industrial empire, the Auergesellschaft of Berlin, which manufactured gas
lamps and electric lights. In 1908, the company asked Haber to
act as technical consultant in its efforts to find new materials for
the filaments inside lightbulbs. It supplied Haber's laboratory with
a variety of hard-to-find compounds, including the rare metal osmium, found in tiny amounts in the earth. Haber promptly began
trying them out as ammonia catalysts.

The third week of March 1909 brought the miracle. Haber
laid a sheet of osmium in his pressure chamber. Once again, he
pumped it full of the usual nitrogen/hydrogen mixture, squeezing
and heating it to the limits of his steel equipment. And finally, like
a rusty lock yielding to a newly discovered key, the stubborn triple
bonds of nitrogen began to crack. Nitrogen atoms joined with hydrogen; ammonia in unprecedented quantity—6 percent of the
total gas mixture—emerged from the reaction chamber. When
chilled to temperatures below freezing, the ammonia turned into
liquid and ran through a narrow tube into a container below.

Haber rushed from one laboratory to another, gathering witnesses to the marvel; he ran upstairs to find Hermann Staudinger,
head of the organic chemistry section. "Come on down! There's
ammonia!" he announced.

"I can still see it," Staudinger said nearly fifty years later.

"There was about a cubic centimeter of ammonia. Then Engler joined us. It was fantastic."

Haber immediately dictated a report to the BASF, a document of understated euphoria. He suggested that the company quietly buy up the world's entire known supply of osmium, most of which lay in the hands of the Auergesellschaft.

Haber waited for the accolades to pour in, but to his surprise and dismay none did. The man in charge of the BASF's research, August Bernthsen, told Haber bluntly that the company had "no interest" in the discovery. Haber's process required outlandish extremes of heat and pressure that would blow up any commercial-scale reaction chamber.

Once again, Carl Engler, the head of the university's chemistry department, stepped in to smooth things over. Engler went over the head of the BASF's research director and talked directly to the chairman of the company's board, Heinrich von Brunck. And as if by magic, Brunck himself appeared in Haber's laboratory accompanied by Bernthsen, the nay-saying head of research, and a younger scientist named Carl Bosch—the same Carl Bosch who had debunked Wilhelm Ostwald's claim of ammonia production nearly a decade earlier.

The three men inspected Haber's experimental equipment. Bernthsen, ever the skeptic, wanted to know how much pressure Haber used. When Haber replied that the reaction needed "at least 100 atmospheres," actually understating the true requirements, Bernthsen professed shock and disbelief: "Why just yesterday only seven atmospheres blew up one of our autoclaves!" Brunck, however, turned to Bosch, who—at least according to the account delivered by Bosch's admiring biographer—saved the day. "I think it can work," said Bosch. "I know exactly what the steel industry can do. It's worth risking." And Brunck decided to follow Bosch's advice. Bosch's confident guess eventually would propel him to the top of Germany's most powerful chemical cartel, I. G. Farben.

Heartened, Haber and Le Rossignol pressed toward a final and convincing demonstration of ammonia-making. Le Rossignol built a new and more efficient apparatus, and Haber discovered a new catalyst—uranium—that was much easier to obtain than osmium and worked just as well.

Haber set the demonstration for July 1–2, 1909. Bosch returned to Karlsruhe with an associate, Alwin Mittasch, ready to see with his own eyes what Haber had promised for so long. But a watched pot won't boil; the apparatus sprang a leak. Bosch didn't want to wait for repairs; he took the next train back to Ludwigshafen.

Mittasch and a BASF technician were thus the company's sole witnesses a few hours later when ammonia began to flow. For five continuous hours, two cubic centimeters of the precious liquid trickled each minute from Le Rossignol's device. Mittasch was "deeply impressed and completely convinced." His report persuaded Bosch to begin all-out efforts to translate Haber's desktop apparatus into factory-scale equipment.

As a practical matter, Haber's work with ammonia was finished. The BASF had adopted his child.

A few days after the successful demonstration to the BASF, Haber organized a celebratory dinner at a local hotel. "He said, 'Invite whomever you like,'" recalled one of Haber's colleagues, the Englishman J. E. Coates. Wine and words flowed freely that night.

Going home afterward, Coates recalled, "we could only walk in a straight line by following the streetcar tracks."

The device that Haber and Le Rossignol built, compact enough to fit on a small table, rests today in the Deutsches Museum in Munich. Sitting quietly, separated from visitors by a short barrier, it's a deceptively modest kernel of a thing, an embryo from which sprouted monsters: machines taller than houses, factories covering hundreds of acres, world wars, and a global flood of grain.

The BASF's Leuna Works, built in 1916–17, for many years the world's largest industrial source of fixed nitrogen.

SEVEN

Myths and Miracles

During the seven plenteous years the earth brought forth abundantly. . . . And Joseph stored up grain in great abundance, like the sand of the sea, until he ceased to measure it, for it could not be measured.

—Genesis 41:47, 49

MANY THOUGHT they knew the meaning of this trickle of ammonia, even during Haber's lifetime. *Brot aus Luft!* they exulted. "Bread out of air!"

The truth, a century later, has turned out to be neither simple nor altogether benign.

Consider a steak. For the sake of simple calculations, assume it's a very large one, weighing twenty ounces. Most of that weight is water. The substances that many care about most—fat and cholesterol—account for only a small fraction of this slab, less than half an ounce at most. Protein makes up much more of it, nearly six ounces. And a sixth of all protein, about one ounce of this steak, is nitrogen.

If we investigate the origins of those nitrogen atoms, and follow their trail, we approach the significance of Haber's invention.

The path of nitrogen leads first to a Midwestern feedlot where one animal among thousands, our steak-to-be, spent its short and brutish life eating a diet rich in protein, and thus rich in nitrogen. From there, the trail splits into many. Streams of nitrogen arrived at this feedlot from thousands of fields across America's celebrated breadbasket sprouting corn and soybeans. Those amber waves of grain fattened our steer and filled it with nitrogen.

The way the story often is told, the trail ends here, as if the fields were inexhaustible, magically transforming sunlight, water, and the sweat of the farmer's brow into food. In reality, those fields, just like our steer, had to be fed on a regular basis. Giant wheeled machines roamed across them several times a year, laying down wide swaths of plant food. The main course of the meal they served was some form of nitrogen: snow-white urea, ammonium nitrate,* or the purest nitrogen fix of all, liquid ammonia injected directly into the soil. Of the crops that fed our steer, corn got the heaviest dose—140 pounds or so of nitrogen on each acre, and sometimes more than 200 pounds. Several valuable crops that humans eat, such as potatoes, grapes, and tomatoes, get fertilized even more heavily.

So we follow this stream of nitrogen further, back toward its

---

*Timothy McVeigh, who built the bomb that destroyed the federal building in Oklahoma City in 1995, and the Irish Republican Army both have used ammonium nitrate as the main explosive ingredient in deadly truck bombs.

source. As we proceed, minor tributaries unite into mighty rivers: processions of barges, rail-borne tanks, and oceangoing vessels bearing enough nitrogen for millions—no, billions and billions— of hamburgers and steaks and cheese pizzas.

We choose the widest and swiftest torrent. It comes from the South, from a place of pipelines and smokestacks, fire, steam, and water.

L ooking westward from the middle of Louisiana's Sunshine Bridge, high above the Mississippi River, a dozen white clouds rise like pillars from the flat landscape ahead. They glow in the light of the rising sun, while the land remains in shadow. The earth, it seems, is erupting, releasing superhuman energy from its depths.

In the shadows lie chemical plants. This strip of Louisiana, sixty miles northwest of New Orleans, is blessed with three things that the chemical industry finds irresistible: plentiful natural gas, tax subsidies, and a great river that leads in one direction toward the open ocean and in the other toward the American heartland.

A few miles to the right, behind a levee that stands guard against the mighty Mississippi, lies the source of a few of those eruptions. It's the Donaldsonville Nitrogen Complex, North America's largest producer of nitrogen fertilizer.

Like all chemical plants, this one is a labyrinth of shining steel pipes, intertwined with a complexity that defies comprehension. Its appetites are enormous. Each day, it consumes a million dollars' worth of natural gas drawn from wells just off Louisiana's shore, in the Gulf of Mexico.

Each day, it also boils thirty thousand tons of river water into steam. The steam, mixed with natural gas, enters the front end of the plant, a giant furnace five stories high. At 1,500 degrees Fahrenheit, steam cracks open the molecules of natural gas, separating hydrogen from carbon.

Then it's on to the next building, another inferno, where air enters the mixture. Oxygen burns away, leaving mostly nitrogen, the other essential ingredient in the recipe for ammonia.

Further chemical reactions strip away carbon dioxide and other gases that would block the union of nitrogen and hydrogen. Three minutes after entering the first furnace, both crucial gases are purified and ready for their "synthesis loop."

A roaring turbine, a cylinder filled with spinning blades, drives the mixture into a vertical steel column about twenty feet tall. This reaction chamber, one of four at the Donaldsonville complex, is filled with beds of iron oxide catalyst, purple-black grains that look like crushed coal. Within these beds, hydrogen and nitrogen combine to form ammonia. Then the gases stream out through a pipe at the bottom of the chamber. They are chilled to below freezing, and ammonia condenses into liquid form.

Underneath the reaction chamber, surrounded by the roar of machinery and the distinctive, unpleasant odor of ammonia, the man who's been leading me through the factory stops and places his hand on a modest silver pump. It is chilly to the touch, and vibrating. He shouts in my ear, his voice barely audible through the din, "All the output!—sixteen hundred tons of ammonia each day!—comes right through this pipe!"

It's a descendant of the fragile trickle that Fritz Haber witnessed in 1909. The chemical reaction is exactly the same, but modern machinery has magnified it a million times.

———

The Donaldsonville complex is a giant nitrogen trap. It sucks up air and converts the air's nitrogen into a form that life on earth can digest.

Its four ammonia production units create more than five thousand tons of ammonia each day, nearly two million tons each year. Some of that ammonia, continually compressed so that it stays a liquid, enters a pipeline that connects this plant with ammonia terminals as far north as South Dakota. Farmers haul it home in white, sausage-shaped tanks. The ammonia can't be sprayed on top of the fields; it would simply evaporate and blow away. Instead, special tractor-drawn equipment slices through the soil, releasing ammonia into the incision, which then closes again around the fertilizer.

Most of the Donaldsonville plant's ammonia production, though, gets converted immediately into chemical forms that farmers can handle with less expensive equipment. Much of it is transformed into snow-white sandy dunes of urea that farmers simply scatter across their fields. In many parts of the world, they do it by hand. Another portion becomes a liquid, white as milk, called urea–ammonia nitrate solution.

Whatever its chemical form, the digestible nitrogen gets loaded aboard gigantic, unwieldy barges on the Mississippi, each one carrying a load equal to 150 railcars. The barges push their way upstream on the Mississippi, heading for fertilizer depots across America's breadbasket, and ultimately farmers' fields.

When the first Donaldsonville ammonia plant began operating in 1966, it represented a new generation of technology and a new era in global agriculture. It was among the first plants to use turbine-driven "centrifugal" compressors, rather than "reciprocating" compressors that work like giant bike pumps. The new compressors were able to move enormous volumes of gas, making possible ammonia production on a previously unthinkable scale. As new plants came online, many of them along Louisiana's stretch of the lower Mississippi, ammonia production in the United States leaped from five million tons a year in 1960 to fourteen million tons in 1970. Visitors came from all over the world to inspect the plants and witness the future.

As ammonia flooded the market, farmers found that they could buy it more cheaply. They also needed more. New lines of hybrid corn delivered larger harvests, but only if the farmers satisfied their larger appetite for nitrogen.

In the 1970s, a shift in government policies added fuel to the engine of agricultural production. Instead of restricting production in order to guarantee stable prices, the Nixon administration advised farmers to grow as much as possible; any excess could be exported. The United States began selling huge quantities of grain to the Soviet Union, and shortages of grain were forecast in other parts of the world. American corn production soared. Farmers in Iowa plowed up pasture and planted crops on that land instead. The land became a grain factory, and it ran on nitrogen. Today, American farmers produce such a mountain of grain that they can barely sell it all.

The revolution spread far beyond the United States. Europe, especially the densely populated Netherlands, became home to some of the most intensively cultivated fields on earth. Even the beloved grasslands of England, where bird-watchers rambled along paths through the countryside, were force-fed a diet of nitrogen. A typical acre of English pasture now receives around 160 pounds of nitrogen annually, as much as North American cornfields.

In South Asia, a "green revolution" took hold, as rice and wheat harvests increased dramatically. Many people in the West assigned sole credit for this astonishing development to newly introduced varieties of wheat and rice. The breeders who created those varieties, though, knew better. One of them, Norman Borlaug, tried to set the record straight upon being awarded the Nobel Peace Prize. "If the high-yielding dwarf wheat and rice varieties are the catalysts that have ignited the green revolution, then chemical fertilizer is the fuel that has powered its forward thrust," he declared in his Nobel Prize acceptance speech.

Worldwide, nearly a hundred million tons of nitrogen are now taken from the air each year, converted into ammonia, and spread across the surface of the earth as fertilizer. Wherever crops are plentiful, from Iowa to India's Punjab, that miracle is possible only because farmers pump unnatural quantities of nitrogen into the soil, retrieving it again in the form of grain or vegetables. The harvests thus achieved are enormous, four times greater than would be possible if they relied solely on the natural recycling of nitrogen. Agriculture—the globe, really—is hooked on the steroidlike drug of nitrogen. According to one careful estimate, about one-third of all the people on earth, about two billion souls, could not survive in the absence of the Haber-Bosch process. Left to its own devices, Earth simply could not grow enough food to feed all six billion of us our accustomed diet.

In July of 1973, readers of the journal *Chemical Engineering* opened their magazines to find a curious article, an account of a business trip to China. It carried the byline "E. J. Mitchell, Warner Engineering," but no such person or company actually existed.

"For business reasons, the company does not wish its identity known at this time," explained the magazine's editor in an accompanying note.

The writer had been among the first American businessmen invited to follow in the footsteps of Richard Nixon, step through the "bamboo curtain," and make a sales pitch in China. His clear and unpretentious prose betrayed the practical mind of an engineer, a man curious about the workings of this unfamiliar place. He described his novel route into China—he walked across the border from Hong Kong—as well as the streets teeming with bicycles, the surprising openness of Chinese negotiators regarding their long-term goals, and the futility of his attempts at humor. "I don't recommend trying to tell jokes in China and don't intend to try it again myself," he wrote. He did not, however, reveal the business opportunity that took him to China, or what he hoped to sell there.

"E. J. Mitchell" was actually James Finneran, an executive from the M. W. Kellogg Company, maker of the world's most efficient ammonia plants of that time. Yet Finneran, despite his careful observation of life around him, never reported on the reason why the Chinese needed his expertise so urgently.

Perhaps he couldn't see it. He ate well during his visit and—at least judging by his report to the readers of *Chemical Engineering*—thought nothing of it. "We ate in ten different restaurants on ten consecutive evenings," he informed his fellow engineers. These meals usually included many courses, thirteen at one particularly luxurious establishment. He and his friends consumed duck, chicken, beef, pork, fish, shrimp, vegetables, fruit, and bread.

Yet beyond the horizon of Finneran's pampered existence in Peking, food shortages gripped eight hundred million Chinese, one-fifth of the world's total population. Seventy years after

William Crookes warned of such a thing, the Chinese—and not the "bread-eating races" of Crookes's imagination—had run headlong into the land's productive limits. Only the ladder of technology—Finneran's ammonia factories—offered a route of escape.

Few outside China have analyzed the country's predicament more closely than Vaclav Smil, a researcher at the University of Winnepeg, in Canada. "China was the most intensive recycler of animal and human waste in the world," Smil says. "They were collecting every little bit of organic matter; children were cutting grasses from slope lands and dumping it on fields. They returned every possible nutrient to the fields, and they'd cultivated every piece of land that could be cultivated. They simply could not produce any more!"

And still it wasn't enough. By the early 1970s, according to Smil, "all food in cities was strictly rationed, meals had to be stir-fried with just teaspoonfuls of precious (and low-quality) cooking oil, and hundreds of millions of peasants subsisted on a monotonous and barely adequate vegetarian diet, eating meat and fish no more than a few times a year." The number of Chinese demanding their daily bowl of rice, meanwhile, continued to increase by ten million or more each year.

Smil has a theory about China's opening to the West in the early 1970s. He's convinced it was driven by the impending food crisis, for within weeks of James Finneran's visit, China placed an order for eight of the biggest and most modern ammonia plants that M. W. Kellogg could offer. Orders for five more plants went to European companies. It was the first large commercial contract between Communist China and its class enemies, a sign of the contract's importance. This also represented the largest single order for fertilizer factories in the history of the chemical industry. More orders followed. Within seven years, China's consumption of nitrogen fertilizer doubled; it became the world's largest consumer of nitrogen fertilizer. By 1989, it vaulted past Russia to become the world's largest manufacturer as well.

Uncounted millions of Chinese farmers spread that nitrogen by hand, using the dry, easily handled form of urea, over tiny plots of land in the coastal provinces of Jiangsu, Zhejiang, Fujian, and Guangdong. These became among the most intensively cultivated fields on the planet, every stalk of rice precious, every cotton bush carefully supplied with fertilizer. On average, each acre of farmland in China's coastal provinces now receives almost three hundred pounds of nitrogen each year. This is more than twice what American farmers put on their cornfields—although the fields in China also produce two harvests each year, compared with a single growing season in North America.

The crops responded, and China experienced the same "green revolution" that had swept across parts of India and Pakistan a few years earlier. As elsewhere, this revolution was ignited by plant breeding, but powered by fertilizer. An acre of Chinese rice paddy now brings forth twice as much grain as it did in 1973. Yields of wheat have tripled over the same period.

China now is home to more than a billion people, and in general, they eat much better than they did a generation ago. Increasingly, Chinese farmers even have the luxury of feeding corn and soybeans to chickens and pigs. And when they savor a bite of that meat, most of the nitrogen atoms in its protein came, originally, from one of China's ammonia factories.

To be sure, the Chinese also are eating better because their factories are humming, churning out exports that pay for grain imported from North and South America. But the entire Chinese economy, including its industrial success, starts with the dramatically rising productivity of its fields. If an army marches on its stomach, as Napoleon supposedly observed, so do armies of urban workers. They're marching on muscles and bones built from industrial nitrogen.

The Chinese story is the piece of Fritz Haber's legacy that fertilizer companies celebrate, a tale of salvation through technology. But it's not the whole nitrogen story, nor the only instructive one.

In the 1960s, Africa was the continent of hope. Unlike China and India, where population bombs ticked toward detonation, Africa faced no apparent Malthusian limits. It had plenty of land to support its population.

Forty years later, however, sub-Saharan Africa has become the new continent of despair. The Asian population bombs fizzled; malnutrition rates among children in India and China fell. In Africa, however, they are now rising. That brutal reality reflects others: poverty, isolation, and exhausted soil.

It's as difficult to settle on the root cause of hunger in Africa as it is to locate the starting point of a circle, and hunger involves circles within circles: Poor people cannot buy food, hunger causes sickness, and sick people become poor; poor farmers cannot buy fertilizer, depleted soils depress harvests, and disappointing harvests make farmers even poorer.

Soil is one piece of the puzzle. In many parts of Africa, William Crookes's bleak vision of nitrogen exhaustion has become reality. On former banana plantations on the shores of Lake Victoria, for instance, fields are depleted of nutrients, unable to pass along any-thing of value to humans. Instead of protein-rich grains, one finds hardy grasses growing there, hardscrabble plants that even cattle prefer not to eat.

For too long, many African farmers in such areas collected their harvests, drawing down the land's nutrient stores. Too many al-lowed rainfall to erode the soil; too few planted nourishing crops of nitrogen-fixing legumes, or plowed crops back into the soil as "green manure." And only a tiny fraction of farmers—mostly those growing cash crops for export—spread fertilizer on their fields. It didn't seem necessary. Land was plentiful. Whenever fields stopped producing good harvests or whenever food ran low, farmers could move on to fresh and more productive land.

Now there is little fresh land left to claim, and much of the land that isn't depleted of nutrients is heading in that direction. Ac-cording to continent-wide surveys, the average acre of land in most countries of sub-Saharan Africa is stripped of at least thirty pounds of essential nutrients—nitrogen, phosphorus, and potash—each year. In some countries, the average depletion comes to more than sixty pounds per acre. Most of this, of course, is nitrogen. It's a slow-moving slide toward ruin.

Experts on the subject disagree about whether depleted soil is a cause of poverty or a consequence of it. Some, a minority, believe that soil depletion isn't even a widespread problem. The reality of stagnating agricultural production and continuing hunger, how-ever, stands uncontested. Meanwhile, a few thousand miles away, warehouses burst with grain and lakes suffer from a deluge of un-wanted nitrogen.

Les Southerland, manager at the Donaldsonville Nitrogen Complex, doesn't think too much about what happens to all that nitrogen after it leaves his plant. When reminded that a portion of it washes from fields and flows down the Mississippi, right past the Donaldsonville plant and on into the Gulf of Mexico, Southerland shrugs. "I don't know about that. We make ammonia here."

But it has to go *somewhere*. In fact, it goes nearly everywhere, spilling through the planet's land, water, and air in a never-ending cascade.

Only a third of the nitrogen that's spread on a cornfield, for example, actually ends up in the corn kernels, a fraction that's typical of other crops as well. Some of the nitrogen atoms disappear into the air as ammonia vapor or nitrous oxide; many of them wash into nearby streams. A few stay in the corn stalk, and so can be plowed back into the soil and serve as fertilizer for the next growing season.

Waste in the field, though, is the least of it. Most of America's grain production goes to feed cattle, chickens, or pigs. When these animals consume soybeans or corn, anywhere from 75 to 95 percent of the nitrogen goes right into their urine and manure. Chicken waste the least nitrogen; cattle the most.

Little of that manure is efficiently recycled. Feedlots and chicken houses often are far away—sometimes thousands of miles away—from fields that need fertilizing. The feedlot operators just want to get that manure off their hands. Too often, the nitrogen washes into nearby streams, leaches into groundwater, or disappears into the air.

Out of every hundred nitrogen atoms that emerge from fertilizer plants like the one in Donaldsonville, only fourteen atoms end up in human food if crops are eaten directly. If the crops are fed to animals instead, with humans eating either milk or meat, only four atoms enter a human mouth. Nor does the trail end there. Humans, just like cattle, excrete most of the nitrogen they consume, and their sewage often gets discharged into rivers and coastal waters.

All of this nitrogen cycles through the environment, from land to water to air and back to land again. Relatively little gets converted by soil bacteria back into its original form—chemically inert pairs of nitrogen atoms in the air. Year after year, as the ammonia factories continue to run, more digestible nitrogen piles up. And everywhere, it causes problems.

A million and a half tons of nitrogen each year float back down the Mississippi toward the Gulf of Mexico, passing barges filled with fertilizer heading the other way. Nitrogen also escapes into the air, in the form of ammonia vapor and nitrogen oxides, at a rate three times higher than was the case a century ago. Cars and power plants release nitrogen oxides as well, but fertilizer remains the major source.

In a few cases, nitrogen pollution affects humans directly. Nitrates pollute groundwater supplies in many farming areas. Nitrogen oxides in the air turn into ozone, harming both vulnerable humans and plant life.

For the most part, humans can survive this outpouring of nitrogen. Wild creatures, though, have a harder time.

Leftover fertilizer is slowly killing streams, lakes, and coastal ecosystems across the northern hemisphere. The changes are gradual, taking place over decades, so it takes a patient eye to notice. Long-term studies, however, reveal dramatic changes.

Fifty years ago, for instance, eelgrass covered most of the Waquoit Bay in Massachusetts. Then came suburban development nearby. Human sewage, containing nitrogen from food taken from a thousand fields, leached into the bay in increasing quantities. Thick beds of seaweed began to grow, crowding out the eelgrass and with it an entire web of natural life, from scallops to small fish.

Hundreds of other waterways around the globe can tell similar and even more dramatic tales. From the Chesapeake Bay to the Baltic Sea and the estuaries of southern China, runaway nutrients from farmers' fields are feeding blooms of algae that cloud the water, suck up oxygen, and suffocate fish. When the fish die, birds that feed on them soon disappear.

When nitrogen oxides in the air come into contact with droplets of water in clouds, nitric acid forms. It returns to earth as acid rain, destroying forests and poisoning streams. At the same time, nitrogen-rich rainfall also fertilizes the land—even land that doesn't need or want fertilizer. Every acre of the Netherlands, whether field or forest, now receives as much nitrogen from rainwater as North American farmers typically apply to their wheat fields on purpose. It's much more than most African farmers could dream of buying. Even smaller doses are enough to play ecological havoc in forests and wild grasslands. Plant species that thrive in the presence of nitrogen start growing uncontrollably, crowding out other plants—and even animals—that aren't used to such conditions. The result is a depleted ecosystem, supporting a less rich and complex web of life.

The surge of nitrogen has become a global environmental phenomenon. It is less famous, but perhaps just as significant, as the "greenhouse effect" caused by burning oil and coal.

Industrial agriculture has its defenders, but it's hard to find anyone who loves it. When it comes to food, we're all romantic agrarians at heart, at least judging from the packaging that lines the aisles of American and European grocery stores. Here we see contented cows munching on grassy pastures and bucolic morning sunrises over pristine fields. There's not a chicken cage, feedlot, or manure lagoon to be seen anywhere, and certainly no tractors spraying fertilizer from giant tanks. Producers of organic food are explicit about this, assuring us that they operate without the aid of "chemical fertilizers."

But human civilization has waded more deeply into the nitrogen flood than most of us realize, and there's no easy way to back out. Most organic producers, for instance, stand only two additional steps downstream from the ammonia factory. They use nitrogen-rich manure as fertilizer, instead of commercial urea or ammonium nitrate. But where did the animals who so generously provided that manure get their nitrogen? From grass and grain, most of which grew on fields made fertile with ammonia-derived fertilizers.

To be sure, organic producers perform a valuable service by recycling that manure. And they contribute less to the nitrogen glut; organic farmers are supposed to minimize their use of external fertilizer altogether by planting nitrogen-fixing crops like alfalfa regularly, or simply by accepting lower, more natural crop yields. But if a conventional corn farmer in Iowa stands hip-deep in the nitrogen river, a grower of organic tomatoes in Fresno County, California, stands at least knee-deep.

Fritz Haber's invention, it seems, has many meanings. It broadened the shoulders of the biosphere, allowing Earth to carry a greater community of life. It also expanded the world's potbelly and clogged its arteries, allowing the people of Europe and North America to consume enormous quantities of meat while polluting the land and water.

For the feeding of humanity, it has been necessary but not sufficient. People have used it to grow more food, but increased food production has surprisingly little to do with the eradication of hunger. Malnutrition and starvation sit cozily, scandalously, beside enormous surpluses of grain, in some cases even within the same country.

Nitrogen fixation teaches a broader lesson: Technology accomplishes nothing by itself. It has no will or moral purpose, any more than the law of gravity does. Human societies create new tools in their own image, and deploy them in the service of their eternal passions. Our machines do not change civilization; like giant mirrors, they reflect it. In the stream of nitrogen, we see humanity's genius, its pursuit of the good life, its inequity, carelessness, and selfishness.

This Law of the Technological Mirror overpowers any inventor's best intentions. William Crookes described nitrogen fixation as the salvation of the "bread-eating races," but it wasn't breadmaking that erected the greatest ammonia factory of Haber's generation. It was war.

The last interpretation of Fritz Haber's accomplishment to surface in public was the earliest and most private. It was a view from the closest perspective possible, that of his wife.

On April 23, 1909, Clara Haber sat at her writing table, her hands hunting in vain for a pen. She silently blamed the men of the house, her husband or her son, for running off with her entire collection of proper writing tools. There was no time to waste; she expected them to return soon, filling the house with noise and distraction. She seized a simple pencil, some black-bordered paper, and began to write: *"Lieber Herr Professor."* It was her last letter to Richard Abegg, her teacher, mentor, and friend.

It was exactly one month after the euphoric day on which Fritz Haber first saw ammonia trickling from his high-pressure apparatus. Ever since, Fritz Haber had been working feverishly in the laboratory, preparing to demonstrate his invention to executives from the BASF.

Clara didn't mention those things in her letter. She probably didn't need to. Abegg had recently visited the Habers, and after his visit, he'd written Clara a postcard. Whatever he wrote in that card—whether congratulations to Fritz or concern about Clara's well-being—broke a dam in her heart. Anger and despair came rushing out onto paper.

> Consider the other side! What Fritz has achieved in these 8 years, I have lost—and even more. And what's left fills me with the deepest dissatisfaction. . . . [E]ven if external circumstances and my own particular temperament are partly to blame for this loss, what's mainly responsible, without a doubt, is Fritz's overwhelming assertion of his own place in the household and in the marriage. It simply destroys any personality that's incapable of asserting itself against him even more ruthlessly. And that's the case with me. And I ask myself if superior intelligence is enough to make one person more valuable than another, and whether aspects of myself that are going to hell because they haven't met the right man aren't more important than the most significant elements of electron the-

ory. . . . Just one more note about Fritz's nature. If I wanted to sacrifice even more of the small life that remains to me here in Karlsruhe, I'd let Fritz desiccate into the most one-sided—though significant—researcher that one could ever imagine. All of Fritz's human qualities, apart from this single one, are nearly shriveled up, and as the expression goes, he's old before his time. On occasion, as in the days we spent in Zurich, a youthful streak still emerges, but anyone who's around him continuously can't escape this impression. He isn't fair enough to look for the reason within himself, but mostly blames me along with our social circle. And last not least, if he weren't kept from it, he'd ruin his health even more than is the case already, despite my truly "harassing" care. Everybody has a right to live their own life, but to nurture one's "quirks" while exhibiting a supreme contempt for everyone else and the most common routines of life—I think that even a genius shouldn't be permitted such behavior, except on a desolate island.

What do you think?

Warm greetings,
Your Clara Haber

Richard Abegg's reply, like the postcard that provoked this letter, has not been preserved. Not quite a year later, in April of 1910, a hot-air balloon that Abegg was piloting crashed, killing him. He was only forty-one years old.

Fritz Haber and Albert Einstein in the Kaiser Wilhelm Institute for Physical Chemistry, a month before the beginning of World War I.

EIGHT

Empire Calls

Leadership in the field of science . . . is of utmost value to the nation politically, not to mention its economic benefits. . . . When [German scientific research] receives the same weapons that foreigners possess, it will be not just equal to them; it will win new peaceful victories!

—Adolf von Harnack,

in a speech to Wilhelm II, 1909

We cannot compete successfully with Germany, in war or in peace, unless we utilize science to the full for military and industrial purposes.

—George Ellery Hale, foreign secretary of the National
Academy of Sciences and founding chairman of the
U.S. National Research Council, 1918

O N MAY 14, 1910, a gaggle of very rich Germans gathered at Wilhelmstrasse 63 in downtown Berlin. They'd been called together by leaders of Prussia, the dominant state in the German empire. The fortunes of these men were built on steel, electrical equipment, chemicals, and banking.

Prussian officials welcomed the moguls, then delivered an extraordinary appeal. The German emperor Wilhelm II, who was also king of Prussia, wanted them to bankroll a new marriage of science and patriotism.

The proposal came originally from the theologian Adolf von Harnack. A few months earlier, Harnack had delivered it in a private speech to the emperor on the state of German science, and Wilhelm had decided to adopt it as his own.

Harnack had depicted the scientific search for knowledge as an arms race among nations, one in which German researchers increasingly were outgunned. In the United States, Rockefeller and Carnegie had established fabulously wealthy research institutes. Like heavy artillery of science, these institutes were equipped to attack problems that lay beyond the grasp of any solitary university professor. Germany possessed nothing similar.

"This cannot, this dare not remain the case, if German science and with it the fatherland—its inward strength and its outward image—are to avoid grave damage," Harnack told the emperor, for a nation's prestige and power rose and fell with its science. As Germany's answer, Harnack proposed, the emperor should establish a series of elite research establishments and name them "Kaiser Wilhelm Institutes."

The German government, however, couldn't pay for them. It

was already spending beyond its means as it tried to expand its navy and overseas territories. This was the reason Wilhelm had called in the industrialists. He was asking them to follow the example of Rockefeller and Carnegie, to donate a portion of their fortunes to this "vital interest" of the German nation.

One man in the room on that day, the reclusive banker Leopold Koppel, listened to the proposal with particular interest. He saw an opportunity to earn the gratitude of the emperor while also creating a scientific powerhouse for his own benefit. And Koppel knew just the man to carry out his plan. Upon returning to his office, he summoned Fritz Haber to Berlin.

I t's a long way from Karlsruhe to Berlin, in more ways than one. When Fritz Haber's train pulled away from the Karlsruhe rail station and turned north into the night, it left behind a small city with modest aspirations and a tradition of political tolerance. When Haber stepped off the train the next morning, he plunged into the noise and bustle of Germany's most unruly metropolis, one where Prussian aristocrats struggled to maintain control over masses of socialist workers.

Haber already knew Leopold Koppel; he'd worked as a technical consultant for one of the companies that Koppel controlled, the Auergesellschaft, which manufactured gas lamps and electric lights. That company, in fact, had been Haber's secret source of osmium, the crucial catalyst in Haber's first successful attempt to make ammonia. Koppel had once offered Haber an extravagant salary to take over the company's in-house research, but Haber had turned down the offer, preferring the freedom and prestige of university life.

From the train station at Friedrichstrasse it was only a short walk to the offices of Koppel and Company on Unter den Linden, the boulevard that ran from the triumphant Brandenburg Gate to the emperor's palace. It's likely that this is where Koppel received Fritz Haber and unveiled his new proposal for the scientist's future.

Koppel laid out the emperor's grand plan for German science, and explained where Haber might fit in. Koppel was ready to finance the construction of an entire Kaiser Wilhelm Institute for physical chemistry and electrochemistry, at a cost of 700,000 marks (roughly $3.5 million in the currency of 2004). He would also cover a large chunk of its annual operating costs, though the institute's director would be employed directly by the Prussian government. He planned to provide this money with only one string attached: that Fritz Haber become the institute's founding director.

Haber was, for Koppel, a kindred spirit. Both were of Jewish background, and both had converted to Christianity. Both had begun their careers as outsiders, yet climbed to the top of their respective fields through hard work and luck. Both had a taste for luxury. The banker's palatial villa in the Tiergarten, a wooded section of central Berlin, boasted an art collection worthy of a museum, with paintings by Rembrandt, Titian, Rubens, and Van Dyck. Unlike Haber, though, Koppel avoided publicity. No photographs of him are known to survive.

The banker's proposal revealed the mind and method of a venture capitalist. He had acquired fabulous wealth by placing canny bets on promising businesses, financing several that had gone on to extraordinary success. Now he was placing a bid to acquire a stake in a promising and ambitious scientist, one who was still rising in his profession and might be genuinely grateful for the aid— unlike, for instance, the already wealthy and prominent Walther

Nernst. Haber wouldn't work directly for Koppel, but Koppel would always enjoy special access to Haber and any particularly useful scientific work done at his institute.

Koppel was disarmingly frank about his own self-interested motives. He preferred to support an institute for physical chemistry, he explained to the Prussian government, "because this field has the most direct impact on the industry to which I am personally closest." In negotiations with the government, he made sure that Haber's institute would be allowed to apply for patents on its discoveries. Koppel probably was hoping that his companies would be able to license those patents and profit from them.

After a whirlwind round of meetings with Koppel and other leading figures in Berlin's scientific establishment, Haber rode home to Karlsruhe, his mind filled with the prospect of a new life in Berlin. Then came months of unexpected silence, as Koppel's brainchild circled within a bureaucratic eddy.

Officials in charge of Prussia's overburdened budget vetoed, at first, the modest new spending demanded by Koppel's institute. Others resisted Koppel's choice of directors; they preferred Haber's archrival, Walther Nernst. Haber, hearing nothing, eventually persuaded himself that life was better in Karlsruhe after all.

Four months later, however, in September of 1910, the Prussian bureaucracy lurched into sudden motion. The hundredth anniversary of Berlin's university was approaching, and the emperor wanted to mark the occasion with a new scientific initiative. Haber received another invitation to meet with officials in Berlin. Upon arrival, he discovered that his future was already determined. The government had scheduled an announcement of Koppel's gift and Haber's new position for the very next day. He had no alternative but to agree, Haber explained later; refusal would have wrecked well-laid imperial plans.

Six weeks later, with the emperor in attendance, the "Kaiser Wilhelm Society for the Advancement of Sciences" rose into life, inflated with the hot air of optimism, ambition, and pride. It represented a unique partnership of imperial sponsorship and private wealth. Private money built the institutes and paid most of their bills, but Prussia paid the salary of each institute director. At the first meeting of the society's governing council, chemist Emil Fischer proclaimed science "the true land of unlimited possibilities." When Fischer mentioned, as one example of science's gifts, the capture of nitrogen fertilizer from the air, he saw the emperor nod his head in agreement.

It took nine additional months for Haber to untie the knots that bound him to Karlsruhe, to extricate himself from his professor's chair and from commitments to various companies. During long visits to Berlin, he oversaw the creation of plans, both architectural and legal, for his new institute. Progress was painfully slow. Haber described his institute as a "giant vat of sausage into which you throw untold amounts of time and effort without yet knowing what exactly will come out of it."

In July of 1911, Fritz, Clara, and their son, Hermann, left behind the city in southern Germany that had been their home for a decade. One of Haber's colleagues in Karlsruhe drew a caricature of Haber's departure. It depicted Fritz Haber in the clutches of an airborne eagle, symbol of the empire. Haber, one arm gripped in the eagle's beak, the other by a claw, appeared helpless. The eagle was carrying him toward an unknown fate.

I n Berlin, wrote Haber's friend, the chemist Richard Willstätter, "Fritz Haber completed his transformation from great researcher to great German." Willstätter's sentence silently acknowledges what Haber gave up in the process. His days at the laboratory bench were now numbered. Never again would his life be fully devoted to personally exploring mysteries of the material world.

The reason lay partly in the sheer logistical challenge of building and running his institute, hiring scientists, and managing its budget. In addition, the field of physical chemistry was changing, and the torch of research was passing to a new generation. Physicists had drawn a new picture of the atom, the basic unit of chemistry, using tools of mathematics that Haber admired, but never fully mastered.

Yet Haber wasn't pulled from the laboratory against his will. He left it voluntarily, if somewhat sadly, drawn by Berlin's sirens of power, honor, and national influence. He sometimes acknowledged the price he'd paid and complained about the distractions of his new role. But whether for reasons of duty or pride, he also relished it. And he never seriously considered leaving it.

Haber accepted—and sometimes demanded—a place in the councils of government, advising Germany's leaders on matters of science, industry, and even war. Head held high, cigar in hand, he strode along Berlin's avenue of power, the Wilhelmstrasse, with the confident assurance of a man who believed he'd reached his proper and appointed station in life.

This, along with his sometimes pompous style, set him apart

from many of his scientific colleagues. "The only deplorable thing about Haber was that he was a bit power-hungry," said James Franck, a young researcher at Haber's institute and future Nobel laureate. "He knew his own intelligence and wanted power. He knew what he was capable of, and his fingers were itching to do it."

He cleared a path into the nation's governing circles through intellect and force of personality. He was named a *Geheimrat*, or privy councillor, a title originating in feudal times that suggested a role as adviser to the throne. By Haber's time it came with no clear responsibilities but still carried great prestige. From this point on, subordinates would no longer address him as *Herr Professor*, but instead as *Herr Geheimrat*. As a member of the nation's elite he also accepted its assumptions: that his nation, surrounded by enemies, demanded his loyalty; that its growing military might serve the cause of peace; that the nation, if united, could never be defeated.

F ritz Haber arrived in a city seized by a fever of imperialism and the adrenaline rush of possible war. Just days before Haber moved to Berlin, a German gunship, the *Panther*, materialized just off the coast of French-dominated Morocco, at the port of Agadir. It was a military provocation, an audacious attempt to intimidate France into turning colonial territories over to Germany.

The "*Panther*'s leap," as the Moroccan crisis came to be called, was only the latest manifestation of Germany's lunge toward empire. A generation of German intellectuals had come to believe that economic expansion demanded territorial expansion; that Germany had a right to its own "place in the sun"—a phrase that expressed a yearning for national honor as well as tropical colonies. University professors preached the "scientific" necessity of empire, influential organizations like the Navy League and the Pan-Germany League promoted it, and many government officials—Wilhelm II in particular—accepted it.

Germany's imperial appetites had given birth to Europe's second largest navy, a railway line toward Baghdad, a few territorial claims in Asia and Africa, and countless conflicts with the established colonial powers of Britain and France. Those adventures destroyed much of the goodwill that Germany previously enjoyed in many foreign capitals, particularly London, yet most Germans remained oblivious to their country's growing political isolation.

During the summer and fall of 1911, as the standoff in Morocco continued, the cheerleaders of imperialism beat the drums of conquest with the slogan "Western Morocco is German!" The French, however, were unimpressed. In the negotiations that settled the crisis, German chancellor Bethmann-Hollweg acquired only economic rights in Morocco and a piece of the Congo.

In Germany, outrage welled up. Shouts of derision greeted the chancellor when he presented the agreement to the Reichstag, Germany's relatively weak parliament. Some suggested that it was time for war, calling for a government that "wouldn't let the good German sword rust."

Bethmann responded with uncharacteristic fury, accusing the nationalist parties of heating "national passions to the boiling point" in pursuit of utopian goals. This was not patriotism, Bethmann raged, but a degradation of patriotism, "throwing away national treasures" of peace and security for partisan political purposes. Only the socialists, normally Bethmann's fiercest opponents, voiced full-throated support for the chancellor's words.

"It is false that in Germany the nation is peaceful and the government is bellicose—the exact opposite is true," noted France's ambassador in Berlin, who was watching the scene with understandably keen interest. The German socialist August Bebel, a fierce opponent of Germany's imperial ambitions and arms buildup, was moved to prophecy:

> So now there will be armament and rearmament on all sides until someday one or the other side will say: Rather an end in horror than horror without end. . . . They will also say: Listen, if we wait longer, we shall be the weaker side instead of the stronger. Then comes the catastrophe. . . . The twilight of the gods of the bourgeois world is in prospect.

Many years later, when historians began trying to explain Germany's imperial obsession, they discovered a curious contradiction. The German intellectuals who pushed most fervently for territorial expansion during this period usually cited Germany's economic appetites—its need for cheap raw materials and export markets. Yet big industrialists, the very people who should have felt these economic hunger pangs most keenly, seemed curiously uninterested in the whole enterprise. They were already enjoying the fruits of an economic boom, and the absence of foreign colonies seemed no handicap whatsoever.

The steel magnate Friedrich Krupp, for instance, saw no need to expel France from Morocco; the French were his partners in a venture to acquire iron ore there. Other German industrialists also counseled patience. In the fall of 1911, one of them confronted the Pan-German League's firebrand leader, Heinrich Class, urging

Class to moderate his warlike stance. "Give us three or four more years of peaceful progress," said Hugo Stinnes, who reigned over vast coal and power-generation holdings, "and Germany will be the undisputed economic master of Europe." Class, who believed that war was inevitable, was dismayed by Stinnes's "delusions."

Unconvinced by economic explanations for German imperialism, historians developed political and even psychological theories instead. They unearthed evidence that Germany's conservative ruling class, led by Wilhelm II, seized upon demonstrations of national "strength" as a tactic to rally popular support against a rising tide of socialist sentiment and cling to power. Instead of concrete political reforms or shorter working hours, the empire offered its citizens intangible participation in national greatness.

The *Panther* steamed toward Agadir in 1911, in fact, partly in search of a propaganda coup that might aid the government in upcoming elections. In this regard the mission failed; the socialists took 34 percent of all votes in the 1912 elections, more than any other party.

Historian Joachim Radkau, meanwhile, proposed a purely subjective basis for Germany's aggressive tendencies—a kind of national neurosis of fragmentation and uncertainty, born of the unsettling social changes in German society. People who yearned for a lost sense of unity and purpose found themselves drawn uneasily toward crisis. According to Radkau, war "had an almost magical attractive power."

Many people of this era did, in fact, describe war as a kind of cure for society's "lethargy and emasculation." Radical nationalists spoke approvingly of war's "strenthening bath of steel." An aide to Chancellor Bethmann-Hollweg may have been hinting at such feelings when he noted, as one factor driving the nation into the Morocco crisis, "the authentically German, idealistic belief that the German people need a war."

Whatever the underlying reason, talk of war grew increasingly common. Some declared it inevitable. Wilhelm himself came to believe that "Russia is systematically preparing for war against us." Fear—a feeling of being encircled by enemies—gave birth to aggression.

Germany certainly was not the sole architect of the disaster that lay ahead. When it came to territorial ambitions in other parts of the globe, England and France had led the way. Britain, intent on preserving its global dominance, did its part to eliminate the possibility of any compromise that would have given Germany a greater global role. But even though every nation in Europe marched toward war in its own characteristic style, Germany marched with an extra swagger in its step.

No other European nation placed such exaggerated confidence in arms as a means of achieving national goals, and none deferred so completely to the presumed wisdom of its military leaders. Bethmann-Hollweg recognized this trait, and worried about it: "We are a young people and perhaps still have too much naive faith in violence, underestimate the finer means, and do not yet understand that what force conquers, force alone will never hold." No other European nation, except perhaps Britain, was so prone to unilateral action, disregarding the interests and sensitivities of its neighbors. Germany's eyes were fixed inward, on its own economic and military ascendance, willfully blind to realities of the world outside.

———

Fritz Haber was no radical nationalist, thirsting for blood. Nor did he, like some Germans, believe in the mystic superiority of Germany's spirit, its *Kultur*. His report on science in the United States in 1903 was devoid of narrow-minded chauvinism, as were his friendly relations with scientists from France, England, and the United States. He was quite capable of amusing himself over the jingoism of his fellow citizens.

"Haber always kept a critical sense of proportion," wrote Haber's friend Rudolf Stern, who later emigrated to the United States. Stern recalled a tense moment in 1913 in a restaurant in Vienna when one Austrian professor "made some biting remarks about the latest rhetorical performances of Wilhelm II." A German lawyer at the table rose to his feet in protest, saying "I cannot abide criticism of His Majesty, particularly not if brought forward in a foreign country." Haber, on the other hand, bridged the awkward moment with an "ironic smile" and some friendly words to the man from Vienna.

For all that, Haber also stood true to Kaiser and *Reich*. For him, there wasn't even a choice in the matter. Duty was its own moral imperative. As a German citizen and a member of the imperial elite, the state's goals defined his own.

The roots of this mentality went deep, all the way back to the traditional allegiance of feudal lords to their monarch. But German intellectuals of the nineteenth century had given it new life and influence, and Haber had, perhaps unconsciously, adopted their view as his own.

These philosophers of state power preached a Teutonic version of "united we stand." The needs of the state took precedence over those of any individual, argued the influential historian Heinrich von Treitschke, for state power alone could protect a society and allow it to triumph over its rivals. Treitschke and others warned against the pursuit of individual interests through political parties; such things sapped the nation of strength and common purpose.

The German elite regarded their unity under the emperor, in fact, as a crucial advantage in the struggle with foreign adversaries. According to the conventional wisdom in Germany, parliamentary democracies such as France were inherently corrupt and internally divided. Germans, on the other hand, possessed a leader who could deliver orders and a population that was ready to respond.

Only the socialists stood outside the camp, insisting that German unity was a fiction, that the nation was divided by class, and that the emperor represented not one *Volk*, but the narrow interests of the property-owning class. Such assertions brought veiled— and eventually openly voiced—charges of treason.

The socialist leader Karl Liebknecht turned a skeptical eye toward universities as well. Science was not the instrument of the common good that it claimed to be, Liebknecht argued in one speech to the German parliament. How could it be, when great industrialists financed their own research institutes, while socialists were legally barred from university professorships? "We social democrats hold no illusions. We know that in a class society . . . such an ideal [of academic freedom] simply cannot be realized." Scientific research and the university itself, Liebknecht asserted, had become a tool of the ruling classes.

It's unlikely that Fritz Haber paid any attention to Liebknecht's speech. Even if he had, the accusation probably wouldn't have bothered him. Nation, government, and emperor, in Haber's mind, were one. He was honored and duty-bound to serve them.

Beyond the western edge of Berlin, in the middle of open fields dotted by windmills, a scientific community began to form. A collection of imposing structures thrust into the air, their walls built of stone and gray stucco, their steep roofs capped with distinctive dark tiles. This was Dahlem, the embryo of a hoped-for "German Oxford."

Fritz Haber was in many ways Dahlem's central personality. He was one of just two founding directors, and the only one who truly ruled his institute. The Institute of Chemistry next door was divided between two competing fiefdoms.

Haber placed his stamp on everything he touched. He influenced the buildings' architecture and named Dahlem's streets for pioneering chemists and physicists. When Dahlem's lack of a single establishment for eating or drinking grew intolerable, Haber appealed to Leopold Koppel, and the banker once again came to the rescue, financing a small restaurant within Haber's institute. It became the community's gathering place until a meetingplace called the Harnack House was built in 1929.

Haber persuaded other scientists to join him in Dahlem. Two of them—Richard Willstätter and Albert Einstein—also had a profound influence on Haber himself. The first would become Haber's soul mate and defender, the second his friendly antithesis.

Willstätter, raised in southern Germany but teaching in Switzerland, was the most brilliant organic chemist of his generation. In 1911, as Fritz Haber was preparing to leave Karlsruhe for Berlin, he read Willstätter's analysis of chlorophyll, the green substance in plants that allows them to capture the sun's energy. Haber was transported into raptures of admiration. Pronouncing himself "overwhelmed," Haber wrote to Willstätter that no scientific research of recent years had exhibited such "power, intuition, and persuasiveness."

"If it could only be arranged for you to come to Berlin!" Haber continued in breathless tones. "I'm motivated by the most naked selfishness, for I feel that it could allow me once again to experience all the enthusiasm of earlier years, if I had the good fortune to observe, close up, your research taking shape."

Other scientific leaders in Berlin had the same idea. They tried to persuade Willstätter to lead the Kaiser Wilhelm Institute for Chemistry, next door to Haber's establishment, and Haber became the intermediary between them and Willstätter. During these discussions, friendship grew. Haber tried, with some awkwardness, to put his feelings into words. "Seldom has it happened, at least as an adult, that I've felt drawn toward affection for a professional colleague with whom I wasn't previously acquainted," Haber wrote on July 31, 1911. "It's been even more rare that the other, of the same age, returns that affection. In your case I do sense this rare and fortunate event."

Willstätter finally accepted a position in Dahlem, drawn toward Berlin by a sense of patriotic duty. "What decided the question," he wrote later, "was that I felt myself a German." Haber arranged for Willstätter to acquire land for his home directly across the street from Haber's more imposing villa, and the two men became neighbors.

Willstätter was smaller and quieter than Haber, but proud and stubborn, with an unshakeable sense of his own identity. Chaim

Weizmann, the Zionist leader, called him "modest, unassuming, and retiring in character; he often reminded me of the old-time venerable type of great Jewish Rabbi."

Willstätter was more deeply Jewish than Haber. "My ancestors were Jews," he wrote, as the first sentence of his autobiography. He was also ardently patriotic, perhaps even more so than Haber. One colleague who knew both men called Willstätter an out-and-out nationalist. Willstätter's ancestors had settled along the Rhine centuries earlier, perhaps as early as the Roman Empire. There they had stayed, becoming rabbis, doctors, and teachers.

Haber and Willstätter forged a friendship that endured to the end of their lives. Haber was the engine of the relationship, irrepressible and impetuous. Willstätter was the ballast and rudder, with a calmness and dry sense of humor that brought the stormy Haber back on course.

Sometimes, the two were like mischievous schoolboys, teasing and rambunctious. Haber once destroyed a large and expensive vase in Willstätter's house while practicing the fine art of walking backward, as court protocol demanded one do when leaving the presence of the German emperor. Willstätter never let Haber forget it, nor the occasions when Haber, immersed in verbally working out some scientific problem, caused them to miss appointments, dinners, and train connections.

Haber insisted that Willstätter acquire a dog for safety in the still-isolated settlement. Willstätter claimed later that this nearly put an end to the friendship. "Mrs. Haber felt hurt because my dog Bobbi barked sarcastic remarks (which were justified, anyway) over the fence to her dog Greif. We, meanwhile, were upset because her evil Greif jumped over the fence and administered severe bite wounds to our friendly though loud Bobbi." Haber had his own version of this story. When he went over to Willstätter's house to patch things up with their new neighbors, he and Willstätter ended up enjoying such an extended conversation over wine that they agreed to address each other henceforth with *Du,* the intimate version of "you" used among friends and family, rather than the formal *Sie.* As Haber put it, "It's so much easier to say *'Du Esel'* ['You ass!'] than *'Sie Esel.'*"

Science was the unquestioned heart of their relationship, as it was of their lives. They admired each other's intellectual gifts and treated their work as a noble calling, a mission more significant than faith, politics, or marriage. When Willstätter composed his memoirs, he wasted barely a line on private matters such as marriage or child rearing. Haber, if he'd had the time or inclination to write the story of his own life, probably would have done the same.

The path of life, Haber once said, meandered along the boundary between two opposing impulses, those of reverence and critique. He might have been describing the opposing inclinations of himself and Albert Einstein.

Haber stood firmly on the side of reverence, tradition, and loyalty. Einstein was all critique, disdainful of conventional wisdom and established institutions, scornful of ties to community or nation. As a teenager, he had taken the remarkable step of formally renouncing his German nationality. While Haber possessed a gift for loyal and devoted friendship, at least toward men, Einstein felt truly bound to no one.

Their friendship was a complicated and episodic affair. Haber revered the younger scientist's intellectual genius; Einstein relied on the older man's organizational skills. And despite their differences, they enjoyed each other's company. They spoke a similar language, sparkling with wit and irony. Both were Jewish, but had distanced themselves from Judaism in different ways, Haber by converting, Einstein by renouncing the religion of Judaism while professing loyalty to the Jewish "tribe."

When Haber helped bring Einstein to Dahlem, Einstein was not yet an international celebrity. He had not yet boggled minds with general relativity and exotic ideas such as travel through time. Within his own community of physicists, though, he was already well known for breathtaking and controversial theories about the motion of molecules and the nature of light. The leaders of German physics, such as Walther Nernst and Max Planck, wanted to bring this intellectual jewel home to the *Reich*, perhaps as director of his very own Kaiser Wilhelm Institute for Physics.

As he had done with Willstätter, Haber agreed to act as an intermediary. On one of his regular trips to a Swiss resort, he met with Einstein in Zurich, hoping to learn what might attract the young genius to Berlin. He also helped loosen Leopold Koppel's purse strings; the banker offered to cover half of Einstein's salary for twelve years.

Following up on that initiative, Nernst and Planck took up the chase personally in the summer of 1913, traveling to Zurich to make the thirty-four-year-old Einstein an extraordinary offer: In addition to his own Kaiser Wilhelm Institute, Einstein could have complete independence in research, a professorship at Berlin's university without any teaching duties, and an exceedingly generous salary. The two illustrious scientists, Einstein remarked afterward, acted like "people who wanted to get hold of a rare postage stamp."

Einstein accepted the offer. He thanked Nernst and Planck for the "colossal honor" they'd shown him. He looked forward with especially keen anticipation, he said, to the chance to work with Max Planck and Fritz Haber.

Privately, he wondered how he would react to Berlin, and how Berlin would react to him. In his rejection of national and religious loyalties, pomp, and social convention, he was the antithesis of official Berlin. He was about to confront the beast that he'd so carelessly taunted.

Albert Einstein's image has softened over the years. Today, many know him only as the gentle and charming eccentric who lived his final years in Princeton, New Jersey. The Einstein of Zurich and Berlin, on the other hand, was a sarcastic intellectual elitist. He was enormously self-centered, a bit of a rake, with an independent streak bright enough to light up the sky.

In a letter to his cousin, with whom he was also having an affair, Einstein composed this devastating portrait of his new friend:

> Haber's picture unfortunately is to be seen everywhere. It pains me every time I think of it. Unfortunately, I have to accept that this otherwise so splendid man has succumbed to personal vanity and not even of the most tasteful kind. This defect is in fact generally and unfortunately a Berlin kind. When these people are together with French or English people, what a difference! how raw and primitive they are. Vanity without authentic self-esteem. Civilization (nicely brushed teeth, elegant ties, dapper snout, perfect suit) but no personal culture (raw in speech, movement, voice, feeling).

Haber, for his part, got a close look at Einstein's own failings, such as his disastrous family life. Shortly after arriving in Berlin in

1914, Einstein began to set conditions for staying married to his wife, Mileva. Eventually, he informed Mileva that their marriage could only continue as a "business relationship." Affection, he told his wife, was impossible. He would behave toward her as he would to a stranger. Mileva initially submitted to Albert's demands, but eventually cracked and decided to move back to Switzerland.

Both Albert and Mileva confided in Fritz Haber throughout their domestic drama, and he seems to have played the role of mediator between them. When the separation became official, it was Fritz Haber who drafted the agreement under which Albert agreed to support Mileva financially. And on July 29, Fritz Haber accompanied the entire Einstein family to the railway station, standing beside a weeping Albert as Mileva and their sons left for Switzerland. Einstein spent that night in Haber's home.

Haber's own home life, of course, was no picture of bliss either. The move to Berlin hadn't eased Clara Haber's unhappiness; if anything, Clara had even fewer friends there than in Karlsruhe. She had little interest in Berlin's glitz and pomp, and Fritz's obvious affinity for such things only deepened the chasm between them.

Perhaps the common domestic misery of Haber and Einstein drew the two men together. Despite all their political and personal differences, Fritz Haber and Albert Einstein would remain close, their destinies linked, for years to come.

By the time Albert Einstein said good-bye to Mileva, the rumble of an approaching avalanche was audible. It had been shaken loose by a volley of shots in Sarajevo. Archduke Ferdinand, heir to Austria's royal throne, was dead from an assassin's bullet. The mass of pride, fear, and national rivalries that had built up over Europe rumbled into motion, gathering destructive force.

A century later, it may seem inexplicable that nations should go to war, and tens of millions die, over one act of terror, the killing of a young man and his wife. In an age of monarchies, however, this assassination carried a powerful symbolic shock, perhaps even comparable to the destruction of New York's World Trade Center.

Archduke Ferdinand became in death the symbol of a nation's pride, greatness, and identity. His killing could be considered no ordinary crime, to be investigated by police, for behind the assassin stood a shadowy network of conspirators, and behind the conspirators stood—at least according to Austria's leaders—the rogue state of Serbia. Austria demanded revenge.

The two grand alliances of Europe shuffled into formation. Germany linked arms with Austria. Russia, supported by France and indirectly by Great Britain, mobilized its military in defense of Serbia. Across Europe, men began unpacking rifles.

Germany's military leaders pressed for a quick decision. They had only one plan for war, and that plan assumed a preemptive attack against foes on all sides, freeing the country from its nightmare of encirclement in one swift strike. The German High Command had planned an all-out assault against France, marching through Belgium in order to bypass the most formidable French fortifications. Then, following a quick victory in the west over France, Germany's generals hoped, they would wheel back toward their other foe, Russia. Absolute victory, in narrow military terms, was the plan's sole goal. There was no room for delay. Nor was there room for negotiation or compromise. There was no plan, for instance, that would allow Germany to defend Austria in the east while maintaining peace with France.

On July 28, 1914, one month after the archduke was assassinated, Fritz Haber dispatched a note to Prussian officials giving notice that he planned to leave, as usual, on a six-week vacation in August. He planned to visit the resort town of Karlsbad, in Bohemia, on the western edge of the Austrian empire.

"Should the political situation develop in such a way that our nation is drawn into entanglements of war," Haber added, he would return immediately to Berlin.

Haber never left for Karlsbad. On August 1, Germany declared war on Russia, and two days later on France. Chancellor Bethmann-Hollweg called it "a leap into the dark."

Fritz Haber gesturing toward chemical munitions during World War I.

NINE

"The Greatest Period
of His Life"

I was one of the mightiest men in Germany. I was more than a great army commander, more than a captain of industry. I was the founder of industries; my work was essential for the economic and military expansion of Germany. All doors were open to me.

—Fritz Haber,
as quoted by Chaim Weizmann

The one phase of Prussianism, borrowed under the stress of war from the enemy, which is likely permanently to remain, is systematic utilization of the scientific expert.

—Thomas Dewey, 1918

IN THE TWILIGHT hours of August 1, 1914, the avalanche of events swept into Berlin. Tens of thousands of well-dressed citizens streamed toward the plaza facing Wilhelm II's palace. Hundreds of miles to the east, Russian soldiers were on the march, preparing for war. Crowds of youth, by all appearances university students, marched along Unter den Linden singing, *"Deutschland, Deutschland, über alles."*

The kaiser and his wife appeared at a window of the palace, and a hush fell over the crowd. "I see no more parties in my *Volk*," cried Wilhelm II. (*Volk*, a linguistic cousin of "folk," sometimes is difficult to translate; it means "the people," but as a collective unity, not a random collection of individuals.) "Among us there are only Germans. . . . All that now matters is that we stand together like brothers, and then God will help the German sword to victory." The crowd cheered.

Many of those present described a holy, transcendent spirit in that place, "an indescribable wonderful unity of sacrifice, brotherhood, belief, and certainty of victory."

One journalist described the scene in these words:

> How the joy glowed through the streets. . . . Now the enthusiasm of the youth has become the joy of men. . . . Total strangers shook hands. The deeper bond of all that is German broke through all the layers of class, ideological, and party differences. Kaiser and *Volk*, government and citizens—all were one.

It wasn't true, of course. It was an illusion, a figment of wishful thinking and crowd psychology. The people who massed around the palace on that day were those most inclined toward patriotic display. In other places, from small towns of the countryside to working-class cities of the Rhineland, and even in neighborhoods of Berlin just a short walk from the palace, observers reported tears, sad faces, quiet and fearful talk, and a "deadening seriousness."

Fear, however, served no great and inspiring purpose, so writers and political leaders focused instead on the crowds around the palace. Mythmakers of the nation, they wished this mystical national unity into existence. For a generation to come, German intellectuals hammered the "Spirit of 1914" into the nation's collective memory until it became part of the definition of German identity.

That spirit is the distilled essence of patriotism, an impulse that certainly is not limited to Germany in 1914. It has erupted in every corner of the world, especially at times of war. It exalts the claims of the nation over all other human ties. As one German speaker put it in 1914: "The *Volk* has risen up as the only thing which has value and which will last. Over all individual fates stands that which we feel as the highest reality: The experience of belonging together."

The full story of Haber's activities during World War I will never be told, for the records have disappeared. Based on the fragmentary evidence that survives, however, he seemed to be everywhere: in his Dahlem laboratory; in the meeting rooms of government offices near the palace; and at military firing ranges and battlefields from Belgium to Poland.

During the very first week of war, as German troops occupied the Belgian city of Liège, Fritz Haber, at age forty-five, volunteered for military duty. Two weeks later, on the day that German armies entered Brussels, Haber appears on a list of industrialists and scientists assembled at the War Ministry, all of them looking desperately for a way to supply the hungry German war machine with ammunition and fuel.

This was Haber's first great contribution to the war effort. Incredible as it may seem, the German military had marched into battle in a state of childlike ignorance about where its own weaponry would come from. The generals—in this respect similar to their counterparts in France, England, and the United States—had accepted the gifts of modern industry without fully comprehending their demands.

They had never bothered to ponder the inconvenient fact, for instance, that the gunpowder propelling every bullet and the explosive charge inside every cannon shell were made from nitric acid, and that nitric acid was manufactured from nitrate that arrived on ships from Chile. The fragile supply lines across the Atlantic now were blocked by British warships.

Germany's military had stockpiled ammunition—but only enough to fight for six months. They were expecting a short war, and quick victory.

The nation's industrialists and scientists, not its generals, first recognized the military peril and leaped into action. A few days after war was declared, Walther Rathenau, leader of the electrical firm AEG, browbeat Germany's top army commander into forming a "raw materials department" within the War Ministry. According to at least one account, some military leaders resisted the initiative because of irritation that civilians would so brazenly interfere in military matters.

Rathenau, a man of imposing intellect and ego, called the world of industry to arms. The leaders of Germany's biggest enterprises, including Carl Bosch from the BASF, Bayer's Carl Duisberg, and the coal magnate Hugo Stinnes, streamed to Rathenau's new home in the War Ministry.

Fritz Haber showed up as well. He had, as yet, no military rank and no factories to his name, but he played a central role in mobilizing the nation's industrial resources. He stood at the intersection of several worlds—military, scientific, and industrial—that had never before worked so closely together. Comfortable in each camp, he pushed them toward a grand alliance. Haber persuaded military officers to adopt new technology, cajoled industrial executives into meeting the government's demands, and assigned scientists the task of solving military problems.

The nitrate crisis overshadowed all other shortages. Since it was, in essence, a nitrogen shortage, Haber's thoughts turned once again toward ammonia.

During the five years after Haber first coaxed ammonia from his high-pressure apparatus, Carl Bosch had become the invention's stepfather, guiding it toward maturity with remarkable speed. A newly constructed BASF factory in the town of Oppau, in southern Germany, already was producing twenty-five tons of ammonia each day. Most of the ammonia was converted into ammonium sulfate and used as fertilizer.

Germany's ammunition factories, however, needed nitrogen in a different form, as nitrate or its chemical cousin, nitric acid. These were the raw materials necessary for making explosives such as nitroglycerine or trinitrotoluene, commonly known as TNT.

There had never been any need for a sturdy chemical bridge between these various nitrogen-bearing compounds. No process existed that would convert large amounts of ammonia (NH_3) into nitric acid (HNO_3). The transformation had been accomplished on a very small scale in the laboratories of the BASF. But no one knew whether it could be carried out in factories, by the ton. Fritz Haber, however, guessed that it could.

There are hints in fragments of surviving correspondence that he began proposing this solution to the nitrate shortage in the very first weeks of the war. On August 21, however, BASF executives balked at the idea, declaring it "impossible" to convert ammonia into nitrate on a large scale. Haber, it appears, was not dissuaded, and kept trying to change Carl Bosch's mind.

In September, just two months into the war, Germany's nitrate shortage suddenly went from worrisome to desperate. At the Marne River east of Paris, French forces mounted a counterattack. The German armies, overextended and exhausted, halted their advance and, in some areas, retreated. Their hopes for a lightning conquest of Paris were extinguished. As the weeks passed, a new and more horrific form of warfare began to emerge on the soggy plains of Flanders.

Just as mass production and coal-driven factories had replaced artisans in small workshops, the machine brought mass destruction to the battlefield. Soldiers became workers in a factory of detonation and death.

Too late, the generals began to grasp the dimensions of Machine Age warfare. Defense trumped offense. Men in trenches, protected from flying bullets by mounds of earth, mowed down onrushing soldiers to the merciless industrial rhythm of machine guns. Commanders facing an entrenched enemy tried to obliterate the trenches by raining down explosive cannon shells on them. In response, soldiers dug deeper into the ground.

The new style of warfare consumed previously unimaginable

quantities of ammunition. Stocks of gunpowder and other explosives that the German military expected to last for a month were consumed in a week. The same quantity soon would be consumed in a day. This wasn't a huge problem for France or Great Britain; those countries were able to import more nitrate from Chile. Germany, on the other hand, could exploit only what it found inside its borders.

Late in September, Rathenau summoned Carl Bosch, the man with his hand on the ammonia spigot, to the War Ministry. There, Bosch learned just how dim Germany's prospects were. Upon returning to BASF headquarters in Ludwigshafen, he called on Alwin Mittasch, the man who'd witnessed the success of Fritz Haber's ammonia experiments.

"Do you think it would be possible to build a factory within a few months that could produce, say, a hundred tons of nitric acid per day?" Bosch wanted to know. Mittasch thought it should be possible. After all, Bosch had already pulled off something similar with the ammonia synthesis.

Bosch contacted army headquarters and delivered his promise. Ammonia from the BASF, he said, could replace Chile's mines and supply the nation with munitions. Work on the first nitrate factory began almost immediately. By May of 1915, it would produce 150 tons of nitrate each day. As the appetite for munitions grew, that factory would be followed by half a dozen more. Haber's ammonia-making process was now feeding the machines of war.

On the fourth of October, 1914, people across Germany opened their newspapers and discovered the following manifesto:

To Civilization!

As representatives of German science and arts, we hereby protest to the civilized world against the lies and slander with which our enemies are endeavoring to stain the honor of Germany in her hard struggle for existence—a struggle that has been forced upon her. . . .

At the bottom of the manifesto stood a long list of famous names in alphabetical order, ninety-three in all. The manifesto's signers were drawn from a broad spectrum of German politics, from dyed-in-the-wool nationalists to liberals who supported democracy and workers' rights. The theater director Max Reinhardt had signed it; so had Max Liebermann, who'd captured the spirit of the German working classes in oil on canvas, and author Gerhard Hauptmann, whose celebration of a nineteenth-century weavers' revolt had infuriated the Prussian aristocracy.

About one-third of the way through the list, before the reader's eye arrived at Walther Nernst, Max Planck, or Richard Willstätter, stood the name Fritz Haber.

It is not true that Germany is guilty of causing this war.
The German people did not want this war. Neither did the government; neither did the Kaiser. . . . Only after overpowering forces that long were lurking on our border fell upon the German people from three sides did they arise as one.

The manifesto was translated (sometimes poorly) into ten languages and rapidly circled the globe. No document of the wartime years provoked greater disgust and outrage among French or

British intellectuals. For decades afterward, some of them would refuse to shake hands with anyone who'd placed his signature on it. The manifesto asserted that Germany was victim, not aggressor. It claimed that the Belgians, not the Germans, had violated civilized behavior with their guerrilla ambushes, while in the east "the earth is saturated with the blood of women and children butchered without mercy by wild Russian troops."

And finally, repudiating those who argued that German society had been hijacked by a military cult, the manifesto delivered a resounding defense of "so-called German militarism."

> Without German militarism, German culture would long since have been extirpated from the face of the earth. . . . The German army and the German people are one. This awareness now unites 70 million Germans, irrespective of education, class, or political party. . . .

This manifesto doesn't sound much like Fritz Haber. Rabid nationalism of this sort doesn't show up in his personal letters.

He was, however, ambitious, driven to claim the honor and success that he felt he deserved. That meant swimming with the social and political tide, riding the current of national unity and purpose that came to be known as the Spirit of 1914. Haber's signature on the manifesto was a sign that he was ready to place his name, his reputation, and his considerable energy at the service of the national cause—a cause that was indeed infused with uncritical nationalism.

German intellectuals, from the safety of their writing desks, roused themselves for a clash of civilizations, a moral crusade in which Germany defended ancient virtues—self-sacrifice, community, and honor—against a modern plague of individualism and hedonism represented by Western democracies. The author Werner Sombart described the international confrontation as a battle between the British "Shopkeeper," possessed by the "narrow, abject spirit of commerce," and the Germanic martial "Hero." In a speech, Sombart declared Germany to be the paramount expression of the "idea of God on earth." Germans, he told his listeners, should pursue the war "with pride, with heads held high, certain in the knowledge that they are the people of God."

Such beliefs were not exceptional. They represented the dominant stream of thought and emotion within Germany. Caught up in the delusion of a moral crusade, Germany, joined increasingly by its foes, could imagine no conclusion other than outright victory. The "spirit of irreconcilability and national arrogance" that was born in August of 1914, writes the historian Wolfgang Mommsen, produced "a deep-seated reluctance to look facts in the face and to pursue policies of realism and moderation."

Despite all the talk of duty and sacrifice, however, Germany still tried to wage war on the cheap. The government didn't raise taxes nearly enough to pay for the war, but instead borrowed money recklessly. It expected to repay the debt with money extracted from its defeated foes after the war.

Haber placed his signature on one more symbolic document eight weeks later, on November 28. He became a member of the newly established German 1914 Society, a social and business club with rooms in downtown Berlin. The club, as its name implied, sought to embody the kaiser's vision of German unity, irrespective of political party. It recruited prominent Germans from across the political spectrum. Many industrialists joined, and Fritz Haber began eating at the club regularly. He would find the contacts he made there extremely useful.

Albert Einstein was among the very few who stood apart, rejecting allegiance to his nation and its young men in battle. In letters to friends he called the war "madness" and blamed Germany's "religious faith in power" for provoking it. Soon after the "Manifesto of the Ninety-three" appeared Einstein signed a countermanifesto calling for European unity and an end to the war. For lack of signatures, however, this document was not released during the war.

Einstein watched with fascination and horror as fellow German scientists, Haber in the lead, laid their skills at the altar of Germany's war efforts. "Our entire much-praised technological progress, and civilization generally," Einstein wrote in 1917, "could be compared to an ax in the hand of a pathological criminal." His friend Fritz Haber, meanwhile—as historian Fritz Stern put it— "began to forge a more powerful ax."

I t didn't take a Nobel laureate to think of using poison gas as a weapon. The idea had been in the air, so to speak, for many years. It took the audacity and energy of Fritz Haber, however, to turn the idea into deadly reality.

Fifteen years earlier, in 1899, delegates to the Hague Peace Conference had imagined the possibility of such weapons and agreed "to abstain from the use of all projectiles the sole object of which is the diffusion of asphyxiating or deleterious gases." (The U.S. delegate declined to endorse this ban, seeing no essential difference between different forms of killing.) Within months of the outbreak of the war, however, small-scale tests of incapacitating gas were under way in Germany, France, and Britain.

In September of 1914, only a month after the war began, an

energetic German officer named Max Bauer proposed that some form of gas be used to drive enemy soldiers from their trenches. General Erich von Falkenhayn, head of the German army, liked the idea and told Bauer to pursue it. Bauer called a meeting to ask for advice from prominent chemists, among them Walther Nernst and Carl Duisberg, head of the chemical giant Bayer. One chemist proposed using dianisidine chlorosulfonate, which produces violent sneezing. The army proceeded to fill a stock of 105-millimeter howitzer shells with this substance, and used them near the end of October near Neuve-Chapelle. The chemical proved useless. It dispersed so rapidly in the air that the French troops never even realized that they'd been subjected to the world's first gas attack.

In Britain, scientists also experimented with tear gas during the fall of 1914. France even went so far as to produce tear gas grenades, possibly intending to use them in combat during the spring of 1915.

An officer in the German War Ministry, meanwhile, remembered his university training in chemistry, during which he'd written a dissertation on benzyl bromides, which sting and burn the eyes of anyone exposed to them. He contacted his brother, a higher-ranking officer who had the authority to order tests of artillery shells filled with one chemical from this family of tear gases, called xylyl bromide.

The tests took place in mid-December 1914 at a firing range near Berlin, and Fritz Haber came to watch. Face-to-face with this novel application of chemistry, probably for the first time, Haber was fascinated. His mind latched onto the problems of gas weaponry and didn't let go for the rest of the war.

Haber loved an intellectual challenge. Here was something new in warfare, its implications still unexplored. His instinctive response was to make the tool work.

The challenge of chemical warfare, in its marriage of the scientific and practical worlds, was the sort at which Haber excelled. Other scientists—Nernst, for example, and above all Albert Einstein—had produced more profound intellectual insights. None, however, possessed Haber's talent for human organization, and these were the skills that war demanded.

When dealing with military matters, Haber learned to "think like a general," said James Franck, a younger scientist who met Haber first in Berlin and later in the trenches of Flanders. At other times, Haber thought like an industrialist, his mind leaping quickly from laboratory experiments to factory-scale feasibility.

But Haber's immersion in wartime work drew on deeper motivations than intellectual aptitude or even his sense of duty. It seemed to fill a psychological need.

"The war years were for Haber the greatest period of his life," wrote Haber's onetime collaborator, the British chemist J. E. Coates, in 1937. "In them he lived and worked on a scale and for a purpose that satisfied his strong urge towards great dramatic vital things. . . . To be a great soldier, to obey and be obeyed—that, as his closest friends knew, was a deep-seated ideal. . . .

"It must not however be supposed that he exalted and enjoyed war *as such*," Coates continued, "but the coming of the war brought out another side of his nature and transformed him into a Prussian officer, autocratic and ruthless in his will to victory."

One of Haber's former students in Karlsruhe, upon meeting his former teacher again during the war years, described a similar transformation. "Haber was not only extremely affable, but—I cannot find any other expression—just fascinating. He is now 100 percent a military man. He believes to have found his true vocation in executing military organizational tasks."

Two decades earlier, as a young man, he'd failed in his efforts to become a Prussian officer. Gas weaponry was a new doorway into the military world from which he'd once been barred. It offered a kind of personal emancipation.

More than half a century after the events of 1915, Haber's youngest son Ludwig Fritz, known as Lutz, a respected economic historian in England, was drawn to examine his father's wartime career. In the masterful study that Lutz Haber finally published in 1986, he draws his own portrait of the father of gas warfare: "In Haber, the OHL [German High Command] found a brilliant mind and an extremely energetic organizer, determined, and possibly unscrupulous."

War had, at first, sucked most of the life from Haber's institute. With so many scientists called to the front, laboratories stood deserted. One wing of Haber's building had been converted into a day care center for children. The scientists who remained soon turned their attention to the military's pressing problems.

By mid-December, days after Haber's visit to the firing range, scientists at Haber's institute began testing new toxic chemicals, hoping to achieve an even more potent cocktail that could fill an artillery shell and make life miserable for French and British soldiers. They turned to some long-forgotten chemicals that chemists previously had avoided because of the exact quality—their toxicity—that now drew their attention.

On December 17, Fritz Haber stood nearby as two of his oldest friends at the institute, Gerhard Just and Otto Sackur, prepared to mix two such chemicals in a test tube. Then someone called from next door. He was needed in the mechanic's shop. Moments after Haber left the room, the test tube erupted in a violent explosion and the laboratory was splattered with blood. The blast blew off Just's hand, but he would survive. Sackur, who'd been looking directly at the mixture, lay dying, horribly mutilated.

Haber came rushing back to the laboratory and, according to one account, collapsed in shock. Speechless, held in a colleague's arms, he could only shake his head as though refusing to believe the scene before him. Clara Haber also came running. She, too, knew Sackur well; long ago, in Breslau, he had been one of two students who'd tested her knowledge during the public awarding of her doctoral degree in chemistry. Clara proved to be "very calm and courageous" in the midst of crisis, ordering the institute's mechanic to cut open Sackur's collar so that he might breathe more easily.

Sackur died shortly thereafter. At the funeral, both Haber and Richard Willstätter were observed weeping uncontrollably.

One day after the accident at Haber's institute, the conception of gas weapons began to take a new turn, away from gases intended to burn a man's eyes and toward something deadlier. Chief of Staff Falkenhayn met with Emil Fischer, professor of chemistry at Berlin's university. Falkenhayn was frustrated. The offensive in the West had stalled.

"He spoke about the new 'stinking materials' and wasn't yet

satisfied with them," Fischer wrote to another chemist after the meeting with Falkenhayn. "He wants something that puts people permanently out of action. I explained to him how hard it is to find materials that are fatal even at very low concentrations. I do know of one very nasty chemical, but I didn't dare recommend it because we don't have the necessary raw materials in Germany to manufacture it. If the enemy should hear about this, it would only hurt us."

Fischer looked and drew back, doubting; Haber leaped. Sometime during the following two weeks Haber approached the High Command with a new suggestion: Clouds of chlorine gas, carried to the front lines in pressurized tanks and released when the wind was favorable, could asphyxiate soldiers in enemy trenches.

The High Command, willing to try anything, put Haber in charge of testing chlorine's effects. A few weeks later, in mid-January, it approved the use of chlorine on the western front, somewhere along the curved line of trenches that surrounded the Belgian city of Ypres.

By February 1915, Fritz Haber, now a scientist-soldier with no rank but great authority, immersed himself in the task of creating a new form of warfare from scratch. He had to recruit soldiers and train them; acquire chlorine and the canisters to hold it. He also had to learn the art of weather prediction, for this weapon would depend on the wind to carry it toward its target.

Haber wanted as many scientists as possible among his "gas troops." He located many who already were serving in the German military elsewhere, and organized their transfers to his unit. He collected an extraordinary group, among them three future

Nobel laureates: James Franck, Gustav Hertz, and Otto Hahn, the future codiscoverer of nuclear fission.

Hahn, a lieutenant in the infantry, was in Brussels when he received an order to meet Haber nearby, in the professor's hotel room. "I came to the hotel, found Haber lying in bed, and from his bed he delivered a lecture about how the war, which had reached a stalemate, had to be waged differently if it was going to be brought to a favorable end," Hahn recalled. "Then he proceeded with a lecture about clouds of chlorine that one would allow to blow over the enemy trenches," forcing those soldiers either to abandon their positions or die.

Hahn wondered aloud whether Germany's use of gas would violate the Hague Convention. Haber told him it would not; that France had already begun using gas—though ineffectually. (This untrue rumor had circulated widely on the German side.) More important, Haber told Hahn, "innumerable human lives would be saved if the war could be ended more quickly in this way."

The motley force of five hundred soldiers assembled in Cologne, where they practiced handling their novel weapons— steel cylinders filled with chlorine under high pressure. Haber instructed the troops himself, making up the rules for chemical combat on the spot. At night, he slept in one of the city's finest hotels, the Domhotel, near Cologne's famous cathedral.

"*Geheimrat* Haber was accompanied by his wife, a nervous lady who was sharply opposed to his accompanying the new gas troops to the front," recalled one member of the group in 1955. But Haber would not be dissuaded. In February, together with his handpicked scientist-soldiers, he packed and drove westward toward the ancient trading city of Ypres.

The land around Ypres, now known by its Flemish name, Ieper, shows the same contours today that it did in 1915. Viewed from a distance, it looks almost flat. Closer inspection reveals a more complicated landscape filled with small dips, gently rising stretches, and a few substantial hills. Little of it is more than 200 or 300 feet above sea level.

Now as then, farmers plow most of this land each spring and plant their crops. If it weren't for the military graveyards that one encounters every few miles, and the occasional bunker—hunks of rough concrete that now lie askew, like misplaced industrial debris—one might easily forget that these quiet acres soaked up more human blood between 1914 and 1918 than almost any other battlefield of the war. Relentless cannon fire blasted away the area's woodlands and erased from view every building in Ypres and its surrounding villages: Langemarck, Poelkapelle, Bikschote, Passchendaele, and many others.

The city still stood intact, however, in March of 1915, when Haber's gas troops arrived at the German front lines five to eight miles outside the city. The war was still in its early stages. Both sides still dreamed of quick victory.

Haber's three companies of engineers, formally under the command of Colonel Otto Peterson, settled in the town of Geluveld, directly east of Ypres. Photographs taken at the time show Haber, potbellied and rumpled, cigar in hand, commanding the attention of all around him, including legitimate officers and casual-looking young scientists in uniform.

Working at night so that French troops in their trenches couldn't observe their preparations—only a hundred feet separated the two sides at some points—Haber's soldiers placed thousands of chlorine cylinders along a ridge called Hill 60 and covered them with dirt. The job proved time-consuming, dangerous, and difficult, for each container was "as heavy and unwieldy as a corpse." The opposing forces, realizing from the unusual amount of activity that something was up, pounded them with cannon shells.

At one point, British troops tunneled under Hill 60 and blew it up, killing scores of German soldiers. The German commander was determined to retake the lost hill and ordered a counterattack. James Franck, the future Nobel laureate, joined in. As the German soldiers reclaimed their lost positions amid a storm of bullets, Franck fell out of sight into a crater. While shells exploded around him, Franck sat calmly at the bottom of the crater collecting samples of air, carrying out his orders. For his exploits, Franck became one of the first gas soldiers to receive a military decoration.

Meanwhile, Haber's weather experts monitored the wind. Most of the time, it seemed to blow the wrong way. Only after all the chlorine cylinders were in place did they obtain weather records showing that this was no passing meteorological phase. They'd chosen the worst conceivable spot for their gas attack. The wind at that location nearly always was in their faces, blowing inland from the coast.

There was nothing to do but move to another location, halfway around the semicircle of German trenches that made up the "Ypres salient." In mid-March, the German gas troops began digging the cylinders out again, moving them to a section of the front about ten miles away, north of Ypres and just to the west of the town of Langemarck. Before they got very far, the meteorologists called a sudden halt: Contrary to all expectations, favorable wind was arriving at the original position.

Troops went on alert. The commander of German forces around Ypres, General Deimling, brought soldiers to the front lines, ready to attack at daybreak on the following morning. At 4 A.M., however, the wind died, and with it any immediate prospect of releasing the chlorine.

General Deimling, thoroughly disgusted with the entire enterprise, rode to Haber's headquarters to give the imported scientist an earful. "I can still see it," recalled one of Haber's soldiers. "How Colonel Peterson and *Geheimrat* Haber stood there, pale and exhausted, and how the general yelled at them. He called them charlatans, accused them of lying to the High Command and lots of other things, too. Colonel Peterson took it all with the stoic calm of a soldier, hand on hat. *Geheimrat* Haber, though, looked very unhappy."

This was not the only occasion when Haber ran into difficulty with the generals. He'd been assigned the Ypres front, in fact, because German commanders in all other sectors refused to accept the new weapons. Some called poison gas "unchivalrous" and others couldn't believe that it would prove practical, especially since the prevailing winds in Europe came from the west.

"I fear it will produce a tremendous scandal in the world," one general wrote to his wife. "I presume [the British-French Entente] soon will have something similarly diabolical. . . . War has nothing to do with chivalry any more. The higher civilization rises, the viler man becomes."

Yet each side in this brutal war was growing more desperate. Even though many military officers remained skeptical of gas weapons to the end, neither side was ready to reject anything that promised quick victory. General Deimling had a change of heart after an enemy artillery shell landed by chance directly on one of the chlorine canisters, destroying it and releasing the gas. By one account, three German soldiers died and fifty were injured. Another observer recalled that twenty soldiers died. "Only then was General Deimling once again convinced of the weapon's terrible effectiveness."

Haber himself got a taste of chlorine on April 2, during a full-scale test of the weapon carried out well behind the front lines. After the cloud of chlorine was released, Haber and an army officer rode too close to the cloud on their horses and nearly suffocated. Haber was sick for several days, but recovered. After the war, he declared that his survival proved that chlorine wasn't nearly as horrible and deadly as the opponents of gas warfare claimed.

"Haber possessed fabulous physical courage," recalled James Franck, who worked beside Haber during those weeks. While observing the test of one bomb, Franck recalled, most of the soldiers threw themselves on the ground before the expected blast. Haber, in his fur coat, remained standing. "He said, 'If I throw myself into the dirt, it'll be much too complicated for me to get up again.'"

By mid-April the cylinders stood ready, waiting for wind from the north. Nerves frayed. Impatient German officers fretted that the British and French might learn of the new weapon. In fact, a captured German soldier did tell his French captors about mysterious containers filled with a deadly gas, but French officers never quite understood the implications.

On April 21, weather experts predicted a reasonably strong wind from the northeast. The German soldiers prepared for attack.

Through the following day they waited. Finally, in the afternoon, the promised wind arrived. At 6 P.M. on April 22, Haber's gas troops opened the valves on five thousand high-pressure steel tanks containing perhaps four hundred tons of chlorine. It hissed through discharge pipes and into the air, then drifted southward toward the French and Canadian lines. A new era had dawned in the uses of chemistry.

From his post in the reserve trenches, well away from the front lines, Canadian Jim Keddie watched, mystified, as yellow-green smoke rose from the German side and drifted in his direction. It formed a wall about fifty feet high and four miles long, moving slowly with the wind, accompanied by the booming thunder of German artillery. At first Keddie thought the eerie cloud was coming directly toward him, but the wind shifted, carrying it further to the east, toward trenches occupied by Algerians who'd been brought to Europe to fight for their French colonial rulers. Keddie caught only a whiff of the gas. "We did not get the full effect of it, but what we did was enough for me," he wrote. "It makes the eyes smart and run. I became violently sick, but this passed off fairly soon."

The Algerians, on the other hand, had no chance. Those who tried to stay in place were quickly overcome, retching and gasping for breath as they died. The rest, if they could, fled in panic, stumbling and falling and tossing away their rifles. The cloud rolled on, moving about a hundred feet each minute. It swept all defense before it, ripping open a four-mile-wide hole in the Allied front.

After about fifteen minutes, the German troops emerged from their trenches and advanced cautiously. Where they previously feared to stand, they now walked with impunity. They clambered through abandoned trenches, barbed-wire and machine-gun emplacements, passing contorted, still-warm bodies along the way. For an hour, they walked unhindered. "Before us lay, still undestroyed, the beautiful old city of Ypres," wrote one soldier. "You could recognize the famous Cloth Hall."

Night, however, was already falling, and the deadly cloud that cleared their way had dissipated. The German soldiers began to dig themselves new trenches, preparing for the counterattack that was sure to come. On the Allied side, reserves were being called up to close the gaping hole in the French lines. Already, they'd thought of some rudimentary defensive equipment: cans of water, along with wads of cotton that soldiers were supposed to soak and hold to their faces. The cotton wads were objectively useless but psychologically comforting, and when the battle began anew the next day, Allied soldiers were never again so taken by surprise and routed by gas. In the end, German forces gained only about a mile of territory through their introduction of gas.

Years after this dramatic "day of Ypres," Haber and other defenders of gas weaponry remained bitter about their inability to take full advantage of it. They felt that an opportunity had been squandered. If only they had been more patient, and waited for a day with good winds in the morning. If only the High Command had provided enough reserves to sustain the offensive, instead of sending them off to the eastern front. Then, German forces might have been able to break through Allied defenses and drive all the way to the Atlantic. The way Haber saw it, the military's tradition-bound lack of faith in his innovation had undermined its potential.

For all that, the events of April 22 shocked each of Europe's armed camps, with far-reaching consequences on both sides. On the German side, there was celebration and long-sought military honors for Fritz Haber. "All at once we and our gas troops became great people," recalled one of Haber's soldiers. "Haber was ordered to appear before the Emperor and promoted from sergeant of the reserves to captain. He appeared proudly in his new uniform, instead of the administrative uniform that we called his 'pest controller's outfit.' " Twenty years after his first attempt, the Jewish merchant's son from Breslau finally had become an officer.

On the Allied side, the outrage went deep. Sir John French, leader of the British forces, condemned Germany's "cynical and barbarous disregard of the well-known usages of civilised war" in a report to his civilian superior, Secretary of State for War Lord Kitchener. "All the scientific resources of Germany have apparently been brought into play to produce a gas of so virulent and poisonous a nature that any human being brought into contact with it is first paralysed and then meets with a lingering and agonising death."

Condemnation, however, proceeded hand in hand with imitation. Within twenty-four hours of the German gas attack, Sir John French wired an urgent demand to London: "Urge that immediate steps be taken to supply similar means of most effective kind for use by our troops. Also essential that our troops should be immediately provided with means of counteracting effect of enemy gases which should be suitable for use when on the move." Some of Britain's leading chemists joined the battle. In the United States, more than 10 percent of the country's chemists eventually would aid the work of the army's Chemical Warfare Service.

The frantic battles around Ypres, during which German forces released chlorine gas four more times, continued for another two weeks. In the midst of the killing, sometime between April 24 and April 29, Fritz Haber returned to Berlin.

It was a quick visit, lasting only until May 2. The conversations, perhaps the altercations, that echoed through the Haber household during those few days are shrouded in mystery. Later accounts by those who claimed to know what happened cannot necessarily be believed, for they weren't present. Those who *were* there, Fritz Haber in particular, refused to speak about it.

This much is sure: On the clear and cool night of May 1–2, under a nearly full moon, Clara Haber found her husband's army-issued pistol, shot herself with it, and died. Fritz Haber, obeying his orders, returned the next day to the front lines of combat. Hermann, just twelve years old, was left behind without mother or father.

The first accounts of Fritz Haber's life, written by his friends, mention Clara's death only in passing. A few ignore it altogether. Richard Willstätter's clipped memory is typical: "Haber was completely a man of duty. I remember the spring day in 1915 on which he returned home for a short visit. It was the day on which his wife died. On that same evening Captain Haber traveled to the eastern front, where he was expected."

As the years passed, and as Haber's work during World War I grew into a symbol of science's uneasy conscience, the interest in Clara's suicide grew as well. Because it came so soon after the gas attacks at Ypres, Clara's final choice became in many minds a symbol of helpless protest, her death a condemnation of her husband's hand in killing.

This apparently was a common view among scientists at Haber's institute. One of them, James Franck, recalled Clara years later as "a good human being who wanted to reform the world. The fact that her husband was involved in gas warfare certainly played a role in her suicide." Franck also felt that Haber "agonized over his guilt in her suicide." A later account of Haber's life asserts, without providing either sources or evidence, that Clara and Fritz argued over gas warfare on the evening before Clara took her life. According to this version, Clara considered gas warfare a "perversion of science."

The most convincing evidence of Clara's opposition to her husband's activities comes from Paul Krassa, a friend of both Fritz and Clara. "A few days before her death she visited my wife," Krassa wrote in 1957. "She was in despair over the horrible consequences of gas warfare, for which she'd seen the preparations, along with the tests on animals."

Yet Krassa couldn't bring himself to draw conclusions about Clara's suicide. "I wouldn't want to decide how much still other circumstances—apart from the conviction that no harmonious life was possible within her marriage—may have played a role in her decision," he wrote. "Can one, or should one speak of guilt in such sad cases? The most heartfelt bonds of friendship united me with both of them, and those bonds remained with Fritz even after Clara's death."

One additional account of Clara's last hours exists, from the hand of the mechanic in Haber's institute, Hermann Lütke. In 1958, when he was contacted by a man hoping to write a biography of Fritz Haber, Lütke dispatched a series of long letters filled with forty-year-old memories of life in Dahlem.

Lütke told a detailed and lurid story, irresistible in its drama. The Habers, he reported, hosted a celebration on the evening of May 1. Among the guests was Charlotte Nathan, the high-spirited young business manager of Haber's downtown club, the German 1914 Society. Late in the evening, Clara came upon her husband and Miss Nathan "in an embarrassing situation" and realized that the two were having an affair. At that point, in Lütke's interpretation, Clara's despair at her fate crossed all tolerable bounds.

The aging institute mechanic recounted the events of the night in exquisite detail: how Fritz went to sleep, sedated as usual with sleeping pills, while Clara sat at her writing desk, composing a suicide note; how she found her husband's revolver and carried it down the stairs from her bedroom to the garden outside; how she raised the weapon, fired a test shot, then turned the weapon on herself. Her son, Hermann, heard the shot and found his dying mother.

Lütke's story, however, was hearsay. He wasn't there himself, but claimed to have learned these details from Haber's household servants. Other friends of Haber do confirm a few parts of Lütke's story—that Clara shot herself with Haber's military revolver, for instance, and that Hermann was the first to arrive at his mother's side. But Lütke also emerges from his letters as a man fond of operatic prose, even when describing less dramatic events. He's not a witness who can be fully trusted.

At the center of the tragedy, meanwhile, Fritz Haber stands inexplicably mute and quickly absent. He left behind no record or memory of any particular acts of mourning, no effort to explain Clara's death, nor any attempt to defend himself against suspicions that he might have driven her to despair. There was no public funeral. Clara was buried quietly and privately in the Dahlem cemetery.

Another death at the same time, remembered much later, helps explain the silence. A week before Clara Haber's death, Fritz Haber's dear friend Richard Willstätter started to worry about his ten-year-old son, Ludwig, who seemed unaccountably tired and thirsty. A doctor examined the youngster and declared him perfectly fit. In reality, Ludwig was suffering from unrecognized diabetes. A day later, he fell into a coma and died.

"It was a time in which a human life meant little," Willstätter wrote years later, his words filled with sad resignation. "On the battlefields of Flanders a generation of German students was being mowed down. On the ever-lengthening front lines the number of the killed and wounded towered into the hundreds of thousands and even higher.

"Once again, as in the wounded days of Zurich [where Willstätter's wife had died suddenly in 1908], daily duties allowed little room for personal things. My recollections of that time and the following months are oddly erased. But the dates of events and of my work allow me to recognize the onward course of my life, now poorer."

In that light, Fritz Haber's rush back to the front lines, into the devouring frenzy of warfare, seems less an act of callousness than of emotional numbness. The world was filled with death. One expression of emotion survives—a letter that Haber sent by military mail from "General Command Post 17" to his former mentor in Karlsruhe, Carl Engler. Haber wrote the letter in June 1915, six weeks after Clara's death.

For a month I doubted that I could keep going. But now the war, with its dreadful images and constant demands on all my powers, has made me calmer. I was fortunate to be able to work at the ministry for eight days, so I could see my son. Now I'm at the front again. Working through all the complications of war with unfamiliar people, I have no time to look left or right, to reflect or sink into my own feelings. The only thing that lives in me is the fear that I won't be able to carry on, or bear the enormous burdens placed on me. . . . It really does me good, every few days, to be at the front, where the bullets fly. There, the only thing that counts is the moment, and the sole duty is whatever one can do within the confines of the trenches. . . . But then it's back to command headquarters, chained to the telephone, and I hear in my heart the words that the poor woman once said, and, in a vision born of weariness, I see her head emerging from between orders and telegrams, and I suffer.

B y mid-1915, Fritz Haber was Germany's czar of gas warfare, his institute a bustling military encampment. Haber commandeered all the empty laboratories he could find, surrounding them with barbed wire and military guards and filling them with a swirl of research on new poisons and gas masks.

"I hope the lion doesn't lay its hand on our modest department," complained Lise Meitner, who continued her more esoteric research into radioactivity in the Institute of Chemistry, next door to Haber's headquarters. "The Haber people treat us of course like conquered territory; they take whatever they want, not what they need."

In 1916, Haber's institute was formally placed under military command. By 1917, Haber's empire encompassed 1,500 people, including 150 scientists, with a budget fifty times larger than the institute's peacetime level. Haber presided over eight separate departments located in various parts of Berlin and a few cities farther away. The actual manufacture of masks and gases took place at industrial factories.

Haber also unleashed poison gas on the lice that plagued German soldiers and the moths that infested flour mills. He followed the example of the United States, where hydrogen cyanide gas was already widely used against insects. The scientists under Haber's command, however, deployed it on a much larger scale. They developed techniques for eradicating insects from entire buildings, including granaries, barracks, trains, prison camps, and warships. The building space was emptied, sealed, then pumped full of hydrogen cyanide.

On the battlefield, clouds of chlorine soon gave way to more toxic chemicals that could be packed inside artillery shells. These weapons didn't require wind to carry the gas toward hostile soldiers. First came phosgene, introduced almost simultaneously by Germany and France. Like chlorine, phosgene asphyxiated its victims. A very different poison, mustard, arrived in 1917, released first by Germany, then by Britain and the United States. Unlike its predecessor chemicals, mustard did not blow away with the wind. It stuck to soil and clothes and wasn't easily washed off. It injured and killed by contact, causing painful blisters on the skin, blinding soldiers who got it in their eyes, and killing those unlucky enough to inhale it. Haber called it a "fabulous success."

From the moment of its first appearance on the battlefield, poison gas has evoked a particular horror, one that Haber, at least, considered irrational. He saw no reason why asphyxiation should be considered more ghastly than, for instance, having one's leg blown off and gradually bleeding to death. And as used during World War I, gas was not conceived as a "weapon of mass destruction." It was not, like the first atomic weapons, unleashed on unprotected civilian populations, though this may be due only to the fact that long-range missiles and heavy bombers hadn't yet been invented.

Yet soldiers seemed more terrified of gas than bullets; perhaps they'd come to accept gunfire as an inevitable part of war's grim lottery. Facing flying steel went hand in hand with bravery, heroism, and manliness. Gas allowed for none of that. It turned a soldier's instincts upside down. The normal direction of safety lay downward: flat against the ground or, better yet, in holes and trenches. Gas, however, being heavier than air, settled into these sanctuaries and turned them into death traps. One was safer standing erect, even mounting the parapets of trenches.

"It is impossible for me to give a real idea of the terror and horror spread among us by this filthy loathsome pestilence," wrote one Canadian officer who faced another cloud of chlorine, somewhat smaller and less panic-inducing, on April 24. "Not, I think, the fear of death or anything supernatural but the great dread that we could not stand the fearful suffocation. . . . Many of the physically strongest men were more affected than their apparently weaker comrades."

The British soldier-poet Wilfred Owen was moved to these words:

> Gas! GAS! Quick boys!—An ecstasy of fumbling,
> Fitting the clumsy helmets just in time;
> But someone still was yelling out and stumbling,

And flound'ring like a man in fire or lime . . .

Dim, through the misty panes and thick green light,

As under a green sea, I saw him drowning.

In all my dreams, before my helpless sight,

He plunges at me, guttering, choking, drowning.

If in some smothering dreams you too could pace

Behind the wagon that we flung him in,

And watch the white eyes writhing in his face,

His hanging face, like a devil's sick of sin;

Come gargling from the froth-corrupted lungs,

Obscene as cancer, bitter as the cud

Of vile, incurable sores on innocent tongues,—

My friend, you would not tell with such high zest

The old Lie: Dulce et decorum est

Pro patria mori.

[It is sweet and proper to die for one's country.]

Such emotional language was foreign to Fritz Haber's mind. He viewed war, and gas in particular, with the cool eye of the technocrat.* He did not, as two British writers later claimed, call chemical weapons a "higher form of killing," but he did think of gas warfare as an intellectual challenge, or an intricate game. Conventional warfare was like checkers, he wrote to the industri-

---

*His writings on chemical warfare, in fact, sound quite familiar to twenty-first-century ears; they strike the same dispassionate tone adopted today by "defense intellectuals" at think tanks such as the RAND Corporation and the International Institute for Strategic Studies. Whether calculating the effectiveness of missile defenses or analyzing the metaphysics of nuclear deterrence, their writings, like Haber's, allow little room for the human horror of warfare.

alist Carl Duisberg. "Gas weapons and gas defense turn warfare into chess."

The heart of this chess game, as Haber saw it, was psychology. Battles were won "not through the physical destruction of the enemy, but rather because of imponderables of the soul that, at the decisive moment, undermine his ability to resist and cause him to imagine defeat," the scientist-turned-strategist told a group of German officers. "These imponderables turn soldiers from a sword in the hand of their leader into a heap of helpless people."

Haber argued that the psychological power of traditional weapons, whether flying bits of metal or ground-shaking artillery shells, was quickly spent. All such "kinetic" weapons were essentially alike, and soldiers got used to them. The bullets might still kill individual soldiers, but they no longer caused the morale of entire armies to crumble.

Chemicals, on the other hand, represented a many-faceted and ever-changing threat. There could be dozens, or even thousands, of lethal chemicals, each one with its own distinctive smell or taste or color—or none at all—and each one requiring a new kind of gas mask filter. Each new poison thus posed a new lethal threat, and a new psychic challenge to the foe, "unsettling the soul." Haber argued that chemicals were thus more powerful militarily than an artillery bombardment, even when they were less deadly. They produced, as he noted enthusiastically in 1925, "more fright and less destruction!"

Gas, the most modern of weapons, naturally worked to the advantage of the most advanced industrial societies, Haber argued. Such countries had the expertise to brew a greater variety of gases, and more lethal ones. Their factories could manufacture more effective masks. And their troops, better educated and trained, were better able to fight while wearing heavy and uncomfortable protective gear. Haber was fascinated by the different technical abilities of Germany's opponents. The British were far better at defense against gas attack than the French, and the Russians were abysmal.

Haber knew well enough that his weapons were widely hated, but he dismissed it the same way he would have dismissed someone's antipathy to the theory of evolution. He saw only one explanation for it: prejudice against anything new and disruptive.

"The disapproval that the knight felt for the man with a gun is repeated by the soldier who shoots steel bullets, when confronted by a man who appears with chemical weapons," he told one audience of military officers.

Those words betray Haber's faith in technical innovation. It wasn't so much a belief that innovation was good—although he did feel, on balance, that it was. Haber wasn't even preoccupied with the morality of innovation; in his view, it was simply inevitable. Whatever *could* be invented *would* be, and there was little use trying to halt or even steer the forward momentum of technology.

Gas weapons, he repeated on many occasions, arose from a kind of technological imperative. Gunfire demanded trenches, and trenches in turn brought forth gas. Whoever mastered this technology best would dominate the battlefield of the future.

Haber's vision, however, was limited to the battlefield. He did not fully comprehend the possibility that future armies would use such weapons—or even more devastating ones—against the civilian populations of towns and cities. Poison gas was not yet a weapon of mass destruction. In this respect, Fritz Haber's imagination remained trapped in the nineteenth century.

Never before had soldiers relied so heavily on the latest products of science and industry. Never before had research institutes worked so intimately with military leaders. Scientists and generals alike began to understand that their once-distant worlds were linked forever.

Gas warfare became one symbol of this union. The continuing nitrogen crisis produced another.

Germany's appetite for ammonia was insatiable. The new factories that converted ammonia into nitric acid were able, at first, to satisfy the military's demand for munitions. But a new shortage emerged at once. With all available nitrogen feeding arms factories, German farmers had no fertilizer. As a result, the harvest was expected to fall by 30 percent, producing food shortages. Adding to the misery was the English blockade on shipments of food, a violation of international law as flagrant as Germany's invasion of Belgium, though less visible. Over the course of the war, an estimated 750,000 Germans died from the effects of hunger.

Near the end of 1914, German leaders hastily created a "nitrate commission." Haber, as usual, was a member of the commission. At its first meeting, he informed the other members that he had a financial interest in the nitrogen question. He didn't, however, disclose precisely how interested he was. Under his contract with the BASF, he was to receive 1.5 pfennig—about one-third of one cent—for every kilogram of ammonia the company produced over the fifteen-year life of Haber's patent on the ammonia-making process. Even before the war started, these royalties had grown to the point where they exceeded Haber's regular salary.

Any decision to buy great quantities of ammonia from the BASF would enrich Haber personally. The other participants, according to the minutes of the meeting, "took notice" of Haber's financial interest but asked for his continued participation.

For the following eighteen months, while also inventing gas warfare, Haber acted as matchmaker for the BASF and Berlin's bureaucracy. He nudged both sides toward a series of business deals that expanded Germany's ammonia production immensely. The deals also made Haber fabulously rich.

In December 1914, the BASF agreed to increase ammonia production at its Oppau factory fourfold, to 37,500 tons per year. Yet the military's hunger for nitric acid, still growing, threatened to consume this increase as well. Within two months, Haber was pleading with the BASF to double production at Oppau once again, to 80,000 tons per year.

The industrialists in Ludwigshafen balked. Carl Bosch worried that he'd be stuck with enormous excess capacity when the war ended. Young men might die for Germany, but Bosch had no interest in risking the financial health of his industrial empire. He demanded that the government pay for any new expansion, while allowing the company a free hand in operating the new plant and setting the price of any ammonia it produced.

A competitor appeared, complicating the negotiations. The industrialist Nikodem Caro offered to produce ammonia using another chemical process, one that was much less efficient because it consumed enormous quantities of electricity. Both industrial teams lobbied officials in Berlin, demanding assurances that would lock in profits. Haber, the BASF's eyes and ears in Berlin, kept the company informed about the state of the government's confidential deliberations.

The ravenous maw of trench warfare, meanwhile, consumed ever-larger mountains of munitions. The sheer enormity of the military's rising demand for nitrogen crushed all resistance to the BASF's negotiating position, for only the Haber-Bosch process was capable of delivering the necessary quantities. In September 1915, the War Ministry asked Haber, a man "who knows how difficult it is to steer the nitrogen ship," to serve as mediator in its acrimonious negotiations with the BASF. In April of 1916, the government gave Carl Bosch what he wanted, and loaned the money for an enormous new ammonia factory. Construction began in central Germany, far away from the threat of Allied air attacks, at the town of Leuna.

The first tank car filled with ammonia left the Leuna Works in the spring of 1917. In 1918, the combined factories of the BASF at Oppau and Leuna churned out 115,000 tons of ammonia, a tenfold increase over the company's prewar production. The ammonia manufactured in 1918 alone was worth 1,725,000 marks in royalty payments to Fritz Haber, the equivalent to about $4 million today. Fritz Haber had become, for his time, an extraordinarily wealthy man.

Scientists abroad marveled at the German marriage of science and warfare, and rushed to imitate it. The United States set up a National Research Council and began a crash program to build nitrate factories of its own. It spent $100 million on them (about $1.4 billion in 2004 dollars) by the end of the war. The project didn't yield much nitrate, but it did forge enduring links between universities and the military. Philosopher John Dewey called this interweaving of science and government policy a kind of borrowed "Prussianism" and predicted that it would remain even after the war had ended.

———

Even soldiers' minds wander. In the midst of war, Fritz Haber found time to fall in love.

Charlotte Nathan was twenty years younger than Fritz, and in many ways the complete opposite of Clara Immerwahr. Charlotte was carefree and impulsive. She had little interest in science, but loved music, dancing, art, and romance. Like Clara, however, she was Jewish.

She worked as business manager at the German 1914 Society, the downtown club that Haber visited almost every day when he was in Berlin. By her own account, this made her "one of the best-paid women in all of Germany." She was comfortable around rich and powerful men, and unintimidated by them. "I received three proposals of marriage in the German Society," she wrote later. "I accepted one."

By her account, they met during the spring of 1917; considering the amount of time Haber spent at the club, though, it's quite likely they met earlier, even if one discounts the gossip of a romantic tryst in Haber's house on the evening before Clara's suicide.

Charlotte wrote that the friendship began when Fritz Haber came to retrieve an umbrella that he thought he'd left behind in her office. The umbrella he found there, unfortunately, wasn't his. With exaggerated gallantry, Haber returned it. "I lay the umbrella on your heart and myself at your feet," he said. Nathan retorted, "I'd prefer the other way around!" and immediately felt embarrassed at her own audacity.

Haber, however, apparently liked her quick wit and returned a few days later for a longer conversation. According to Charlotte's story, which strains credibility, it turned into a three-hour recounting of the most intimate details of Fritz Haber's life, including the misery of his marriage, which he blamed completely on Clara. Ever since his trip to the United States in 1902, he told Charlotte, Clara and he had slept in separate rooms.

Charlotte Nathan wrote that she was overcome by Haber's passion and vulnerability: "A very different man suddenly stood here before me: a man who sought and needed love—love that I could give him. I hesitated, but at the same time felt that we belonged together, with soul and body."

Romantic weekend excursions followed. Within weeks, on Easter weekend of 1917, they were engaged. But months then passed without any move toward marriage. Charlotte's letters grew impatient. Fritz, in reply, expressed doubts.

> . . . I don't love you any less than you love me, but I don't have the inner certainty of what we can become to each other when we commit ourselves to each other forever and live together for a few years, because at that point the love that we feel now will be finished. In its place must emerge an inner harmony in our relationships with other people and the world around us, and we don't know yet whether we possess that harmony.

Haber evidently overcame his hesitations. He soon dispatched a telegraph to Charlotte proclaiming, "Miss Nathan's last journey." Soon, he wrote, she would be "Mrs."

On October 25, 1917, Captain Fritz Haber, outfitted in military uniform complete with sword and spiked helmet of a Prussian officer, took Charlotte Nathan as his wife. The ceremony occurred at the spiritual altar of the German empire, inside Berlin's Kaiser

Wilhelm Memorial Church. According to Charlotte, Haber refused to have the wedding anywhere else, which meant that Charlotte had to convert to Christianity. She had chafed at her husband's demand, which she found unreasonable, but complied. The wedding thus became a symbol of many things: of love and new beginnings, but also of Haber's desire for social status and his loyalty to the national cause.

Within weeks, a child was on its way. Eva-Charlotte was born on July 21, 1918. Conflict in the marriage appeared just as quickly. Like Clara before her, Charlotte resented Fritz's high-handed ways. Unlike Clara, she waged a spirited counterattack. She demanded Fritz's attention, protesting against his extended absences from home and his disinterest in family life while there. Half a year after the wedding, Charlotte complained to Fritz Haber's father Siegfried about her husband's "oppressive" disregard for her own desires.

Fritz, meanwhile, was startled to discover that his wife wasn't happy just minding her own business at home with an adolescent stepson and newborn daughter. He was upset when she announced her plans, on just a few days' notice, to travel to a resort in Switzerland. When Haber became crippled by one of his periodic nervous breakdowns, he blamed his wife, calling her demands "thumbscrews of the soul."

By the summer of 1918, German forces in the field were exhausted and defeated, although fanatics in Berlin and Munich refused to see it. As fresh British tanks crushed German defenses, it became increasingly difficult for ordinary German soldiers to believe in victory. The "imponderables of the soul" that Haber believed were the key to success in battle—along with the industrial might of the United States—had turned decisively against Germany.

There were plenty of Germans—too many—who still believed in mystical forces that somehow would carry Germany to victory. But Haber probably was not one of them.

Sir Harold Hartley, a British expert on chemical weapons who became Haber's friend after the war, enjoyed repeating Haber's war story about a major German offensive planned for 1918. The Germans wanted to rely on gas during this attack, but the wind was blowing in the wrong direction. Then the wind turned, and a relieved Haber went to Field Marshal von Hindenburg and announced, "Field Marshal, the wind will be favorable tomorrow morning in obedience to your orders." Hindenburg didn't recognize Haber's irony. He stood up, raised his right hand, and said, "Not in obedience to my orders but by the will of God."

"He spoke from the heart," Haber told Hartley, and added with a cynical smile, "There certainly are curious people in the world."

Other witnesses recount a meeting in 1917 at the War Ministry with the most powerful German general, Erich Ludendorff, at which Haber announced the availability of a remarkable gas weapon called Yellow Cross, now known as mustard gas. Haber warned Ludendorff against using the new weapon unless he was convinced that the war could be won within one year. For within one year of using Yellow Cross, Haber told the general, Germany's enemies would be able to produce it themselves, and use it against Germany. The new weapon compelled any soldier exposed to it immediately to switch into new clothes. That might be possible for England, Haber warned, but not for resource-starved Germany. Germany did use mustard gas, however, and the British responded with mustard attacks a year later.

According to one German officer, Haber practically lectured Hindenburg and Ludendorff like students at one point, trying to persuade them to end the war as soon as possible under relatively tolerable conditions.

"These are almost his exact words, I will never forget them," recalled the officer. "Your excellency, we can only win the war if we are able to build enough submarines to block enemy reinforcements, and enough aircraft to achieve mastery of the air. But we are fighting a poor man's war. We cannot build enough of both of them—enough submarines *and* enough aircraft."

An interesting aspect of this quote is its fixation, so characteristic of Haber, on technological factors. Yet at the close of the war, Haber was forced to recognize that technology had proven to be not a savior, but a cruel temptress. Germany had pursued a series of *Wunderwaffen*—"miracle weapons"—believing in technology's promise. Yet none of those weapons, from gas weapons to submarines, had changed the course of the war. Having awakened exaggerated hopes, they made defeat all the harder to bear.

In fact, the *Wunderwaffe* that aroused the greatest enthusiasm in Germany, the submarine, actually sealed Germany's defeat. Unrestricted attacks on Atlantic shipping, including civilian ocean liners such as the *Lusitania*, provoked international outrage and brought the United States into the war against Germany. Most historians agree that this was Germany's greatest mistake of the entire war.

Even Haber's most important technological triumph, the ammonia factories that delivered munitions to German soldiers and fertilizer to German farms, is in hindsight of deeply ambiguous significance. Those factories didn't change the outcome of the war, but prolonged it by three years, piling horror on top of horror. That brutality bred more brutality.

If Germany had been forced to surrender in 1915, Lenin might never have made it back to Russia from exile in Switzerland. The Bolshevik revolution might never have happened, or it might have taken a milder course. Germany, too, would have been much less likely to descend into economic chaos and political bloodletting. In the absence of Fritz Haber, in other words, we might never have heard the names Hitler and Stalin.

A t the end of September 1918, Germany's military commanders suddenly declared the war lost and demanded that civilian leaders negotiate a peace settlement. But weary soldiers, sailors, and workers took matters into their own hands. Mutinies broke out in early November, first in port cities, then in other large cities with military bases and factories. Workers' councils occupied factories and took over city halls, demanding an end to the war and the overthrow of Germany's government.

On November 9, 1918, Kaiser Wilhelm II abandoned his throne and went into exile. Friedrich Ebert, the leader of the Social Democratic Party, proclaimed the end of imperial Germany and the start of a new republic. On November 11, all sides signed an armistice, and the guns went quiet. A shell-shocked continent paused to consider its losses.

The war had killed 1.7 million Germans and wounded or disabled another 4 million. This amounted to one-tenth of Germany's entire population at the start of the war. The totals in France, as a percentage of the country's population, were even higher. The total number of killed and wounded in all countries came to about 20 million.

Chemical weapons accounted for a relatively small number of these casualties. On the western front, where gas was most heavily used, it killed or injured about 650,000 people. Gas weapons claimed most of their casualties in the last year of the war, when mustard was introduced. The numbers of casualties from the German use of gas in Poland and Russia are unknown.

A cemetery for Germany's World War I dead stands alongside a country road near Langemarck, a few miles east of the rebuilt city of Ypres. It overlooks, by coincidence, the place where gas warfare was born. The cemetery, which holds the remains of forty-four thousand soldiers, is a shadowy and somber place, very different from the British military cemeteries that dot the landscape nearby. The British grave markers stand proudly upright, white in the bright sun, still celebrating duty, honor, and wartime sacrifice. Most of the German markers, dark in color, lie flat against the ground, shaded by trees. Whereas the main British cemetery at Tyne Cot, a few miles to the east, entertains a steady stream of visitors, the memorial at Langemarck is often deserted. Perhaps the Germans, more than the British, have had the idea of martial celebration beaten out of them.

I stood amid the gravestones during one recent visit and heard an unseen jet scream across the sky. Judging by the sound, it was a military fighter jet on patrol, probably carrying some of the most advanced weaponry in NATO's arsenal. It suddenly struck me: This high-tech aircraft, rather than the chemical arsenals hidden in two dozen or so nations around the world, represents the true legacy of Fritz Haber the gas warrior.

Chemistry and toxic gases represented the technological cutting edge of warfare during World War I. But they don't any longer. Poison gas is old technology. Nearly all the chemical weapons that remain in military arsenals today were invented at least half a century ago.

Chemistry was supplanted during World War II by physics, which brought forth treasures: radar, proximity fuses, electronic computers, and finally the atom bomb. More recently, the elite armies of the world have turned to the fruits of computer science and electronic engineering. They fight wars by remote control, with video cameras aboard pilotless aircraft and bombs that home in precisely on geographic coordinates delivered by orbiting satellites.

John Dewey's prophecy in 1918 has been proven correct; the marriage of science and military power has endured. And its spiritual lineage leads back to Dahlem.

The architects of high-tech combat walk, not realizing it, in the footsteps of Fritz Haber. Theirs is the latest "higher form of killing," the latest dream of technological advantage in war.

Fritz and Charlotte Haber, accompanied by a Japanese friend, in Egypt during their round-the-world trip in 1924.

TEN

Like Fire in the Hands of Children

Only scientific progress can restore all that the war destroyed.

—Fritz Haber, 1923

The great technical accomplishments that the past 50 years have granted us, when controlled by primitive egoists, are like fire in the hands of small children.

—Fritz Haber, 1932

FRITZ HABER TURNED fifty on December 9, 1918, a month after the war ended. There were no great celebrations. He was surrounded by too many worries, and too much exhaustion.

Haber and his nation rose and fell together as though connected by an invisible thread. As Germany's economy crumbled and its political system came apart at the seams, Haber suffered as well. "You know the feeling when you're on a snow-covered slope, sliding downward?" he wrote to the chemical magnate Carl Duisberg in February 1919, two months after the end of the war. "You don't know until you get to the bottom whether you'll arrive with all your limbs intact or with broken legs and neck. All you can do during the slide is stay calm. . . . This mountaineering experience is what we're going through—painfully—in economic life at the moment."

During the first five years of the Weimar Republic, from 1919 through 1923, nothing was secure, everything uncertain. Bands of armed young men brought the war home and turned it inward, brawling with everyone whom they considered responsible for Germany's shameful surrender. For the first time in memory, assassination became a tool of German politics. Inflation turned cash into trash.

For Fritz Haber, too, it was a time of clashing extremes, of honor and dishonor. One day he feared being placed on trial as a war criminal; the next he received science's most prestigious prize. He secretly advised a network of military officers and industrialists who hoped to rebuild Germany's military arsenal, yet maintained friendships with socialists and pacifists.

Many saw him as powerful, but Haber felt increasingly powerless, reduced to "a vase that one places in a room because it seems to fit there, where it stands until someone accidentally breaks it." He was a German patriot who also believed in political tolerance, who tried in his own way to nurse Germany's crippled democracy toward health. As it turned out, neither the man nor the democracy was strong enough to survive for long.

During the Weimar Republic's entire fourteen-year lifespan, Fritz Haber was plagued by depression, physical weakness, and a failing heart. Resignation, however, was not part of his nature, and the widening gap between his dreams and his strength left him increasingly frustrated. "I'm too ambitious," he confessed to his friend Richard Willstätter, the one man to whom Haber poured out his despair. "What I *can* do isn't enough for me, and what I *can't* do, I no longer can learn."

Fritz Haber never considered himself an outlaw, but the aftermath of the war forced him to think like one. During the summer of 1919, there were rumors—perhaps true, perhaps not—that his name stood on a list of war criminals whom the Allies wished to arrest.

The threat became urgent in July. Germany ratified the Treaty of Versailles, which included a provision that compelled it to turn over accused war criminals for trial. Haber immediately sent the rest of his family—Charlotte, seventeen-year-old Hermann, and Eva, who was not quite two—to the neutral haven of Switzerland. Perhaps he acquired a new identity; years later, scientists at his institute discovered a forged passport with Haber's picture. On August 1, he followed his family to the Alpine city of St.-Moritz. There he waited. He also grew a beard, evidently thinking that it might help conceal his identity.

The criminal charges that Haber feared never materialized, and he returned to Berlin a few months later. During his time in Switzerland, however, he encountered moral condemnation of another and more personal sort. It came unexpectedly, in the form of letters from an old acquaintance, the chemist Hermann

Staudinger, who had occupied a laboratory one floor above Haber's at the university in Karlsruhe.

Staudinger had spent the war years in Switzerland, at Zurich's university. He'd turned increasingly against the war; the use of poison gas, in particular, appalled him. By 1917, he had begun writing letters and publishing articles arguing that Germany's defeat was inevitable, and advocating an immediate end to hostilities.

When Staudinger heard that Haber was in Switzerland, he sent his former colleague a letter, including with it several of his wartime publications. Haber sent a brief, dismissive response, suggesting that Staudinger's hostility to chemical weapons was outdated. Haber recommended that Staudinger read a variety of German, English, and American publications that supported the legality of gas warfare.

Staudinger didn't give up. He wrote back, asking Haber to consider not just whether gas weapons were legal, but also whether they were moral. His words to Haber foreshadow the thoughts that emerged half a century later among America's nuclear scientists: "I hoped that you might agree with this view: That we, as chemists, have a special responsibility in the future to point out the dangers of modern technology, and in so doing to promote peaceful relations in Europe, since the devastation of another war would be almost unthinkable."

This time Haber replied at length, and bitterly. Staudinger's views were "divorced from the real world," he wrote, because the banning of particular chemical processes wouldn't end war.

Eternal peace can't be assured through technical means. A husband and wife get along because of their spirit and self-discipline, not because you lock up every rod and poker. Still, this difference of opinion wouldn't estrange me from you. What bothers me is something that I think you don't even see—the real effect, even if unintentional, that your writings had at the time you wrote them. You stabbed Germany in the back in the hour of its greatest need.

At the moment when Staudinger published his articles, Haber continued, Germany's enemies were busy digging up any claim of German atrocities that could be used as grounds for an even more punitive peace settlement. In Haber's conception of right and duty, Staudinger had an obligation to refute such slanders. Instead, he'd made them his own, repeating foreign charges "because you thought they would help support the realization of your pacifistic ideals." In so doing, Haber wrote, Staudinger had betrayed both "truth and homeland."

Staudinger's arguments had caused grave damage to the *Reich,* Haber concluded. His final words to Staudinger cut like daggers: "The damage remains, even if unintentional. It can't be repaired, and that's what separates me from you."

Haber didn't know it, but Staudinger had done more than simply publish antiwar articles. In January of 1918, Staudinger had learned about Germany's plans to use mustard gas that spring. He was so horrified that he warned an American colleague that a new and much more deadly gas weapon was on its way. The American scientist passed the warning along to officials of the International Red Cross, who informed French officials. The warning apparently had no effect on Allied operations. For Staudinger, it was an act of conscience and moral responsibility; Haber undoubtedly would have called it treason.

Haber wasn't angry just at Staudinger; he was angry at the world. He considered himself and his country victims of political persecution. The Versailles Treaty prohibited German chemical weapons, but the ban had no moral or legal legitimacy in Haber's eyes; its unequal terms were rooted only in military power. Britain, France, and the United States had used chemical weapons as well, and they were permitted to improve those arsenals after the war. If the Allies were prepared to seize every advantage that power and circumstance allowed, Haber felt that Germany should do the same. Haber was quite ready to violate the treaty's terms if he could get away with it.

In this respect, Haber represented the first of a breed. He was the forerunner of every modern scientist who works on banned weapons—at least those weapons, such as nuclear bombs, that international treaties allow in a few privileged nations but not in others, such as Iran or North Korea.

When the British chemist Harold Hartley, acting as an international arms inspector, arrived at Haber's institute in 1921 to check for research on forbidden weapons, Haber greeted him with a bit of theater. "Why haven't you come before?" he asked Hartley with mock dismay. "I was looking forward to going over our records with you and only last month we had a most unfortunate fire. They were all burnt. Look at the roof!" Hartley, who enjoyed telling this story for years afterward, looked up and saw a large hole covered with tarpaulins. "I'm sure you have a good memory," Hartley replied.

Haber and Hartley, who had helped direct Britain's chemical warfare efforts, spent an enjoyable week trading war stories. Haber had more in common with the foe Hartley, in fact, than with his fellow countryman Staudinger. Neither man considered poison gas any more sinful than ordinary bullets. "We were soon discussing the pros and cons of gas tactics and defence much as we should have discussed any other scientific problems," Hartley reported. "I like to think that we parted at the end of a fortnight as friends. It was a great experience to have enjoyed his confidence."

Haber, however, kept some secrets from Hartley. It's unlikely that he told the British chemist about the work that a few of his former associates were carrying out at the nearby Imperial Biological Institution for Agriculture and Forestry.

The scientists in question were experts on the chemical control of insects. They'd led the wartime campaign against lice and moth infestations in military facilities. After the war, when Haber's institute dropped all direct research on toxic gases, they moved on. But at their new laboratory, ostensibly devoted to insect control, they built facilities to test toxic chemicals on warm-blooded animals such as mice, rats, rabbits, and guinea pigs. The funding for these facilities came, it was said at the time, from an anonymous "Dutch friend." In reality, the laboratory was funded in part by the German military, arranged through the good offices of Fritz Haber. Not only did this laboratory routinely work with banned chemicals that Germany had once used as weapons; part of its job apparently was to test potential new chemical weapons in violation of the Versailles Treaty.

Haber also didn't tell Hartley about the activities of Hugo Stoltzenberg, an entrepreneur who had worked at Haber's institute during the war. After the war, Haber persuaded Stoltzenberg to take on the dangerous job of cleaning up a partially destroyed gas weapons plant in Breloh, on the plains of northern Germany.

Stoltzenberg cleaned up the site, acquired in the process a considerable stock of chemical weapons, and became an arms

merchant. By 1921, when Hartley came to visit Haber, Stoltzenberg had begun selling chemical weapons to Spain, which used
them to help crush a rebel uprising in Morocco. This led to a contract with the Spanish government to build an entire factory for
production of mustard and phosgene. Fritz Haber arranged some
of Stoltzenberg's contacts in Spain, and it's likely that Haber knew
exactly what his former aide was up to.

A few years later, Haber opened doors for Stoltzenberg's most
daring and ill-advised venture, a gas weapons factory in the Soviet
Union. The idea first took shape in 1923, when the young Communist state approached its equally destitute German neighbor
with a proposal of military cooperation. The Germans had technical expertise; the Soviet Union had plenty of land, natural resources, and the advantage of being unconstrained by treaties
banning controversial weapons.

The German military, looking for someone with the technical
competence to build a secret chemical weapons factory, asked
Haber for advice. Haber suggested Hugo Stoltzenberg. In short
order, Stoltzenberg was commissioned to build a mustard gas factory near Samara, along the Volga River. The factory, however,
turned out to be much more expensive than Stoltzenberg had
anticipated, and two years later he ran out of money. In the meantime the political winds had shifted; Berlin no longer supported
the project, and Moscow no longer wished to finance it. Stoltzenberg was ruined. In 1926, Fritz Haber ended up presiding over
secret negotiations between Stoltzenberg and the German government that settled Stoltzenberg's outstanding debts.

As far as is known, this was the final episode of Haber's work in
the shadowy world of the weapons underground.

Fritz Haber had written off the possibility of honor in Stockholm—or perhaps he just didn't want to jinx his chances. When Richard Willstätter, who'd received the Nobel Prize for chemistry in 1915, nominated Haber for the prize immediately after the war, Haber responded in a letter to his friend that "I find it unbearable to write about such things. It's all so dreary and irrelevant, and I have no doubt that political considerations make it inconceivable for Stockholm to consider Germans who've been recommended by other Germans." Nor was he convinced that his scientific legacy amounted to much: "I always jumped from one thing to another."

Haber's self-criticism was accurate; he knew his own strengths and weaknesses. His mind, quick and sharp, was unparalleled in criticism, analyzing the strong and weak points of any argument. He could communicate scientific ideas, either in lectures or essays, better than almost any other scientist. Those gifts made him a superb teacher, mentor, and leader of research teams. What he lacked was intuition, a sense for what might lie beyond the horizon. His greatest accomplishment, the synthesis of ammonia, was more a product of raw determination and technical skill than intellectual brilliance.

Members of the Nobel Committee in Stockholm, however, felt that Haber's strengths were impressive enough. They'd been discussing his candidacy for several years, and understood the enormous significance of the ammonia synthesis. It nearly always took a few years for any accomplishment to be recognized. Now it was Haber's turn.

When the news arrived in mid-November 1919, Haber seemed happier for his country than for himself. It was, indeed, a great day for German scientists; along with Haber's prize for chemistry, the Royal Swedish Academy of Sciences awarded two prizes in physics, for 1918 and 1919, to the German scientists Max Planck and Johannes Stark. "I think it was a deed of greatness on the part of the Swedish academy to elect three Germans—and only Germans—as prizewinners," Haber wrote to Willstätter. "My heartfelt wish is that it may lead to renewed international understanding."

It led, instead, to an immediate howl of indignation, particularly in Belgium and France. Two Frenchmen who were to receive the Nobel Prizes for medicine and economics rejected their prizes in protest. The *New York Times* wondered scarcastically "why the Nobel prize for idealistic and imaginative literature was not given to the man who wrote General Ludendorff's daily communiqués." Another American, winner of the Nobel Prize for chemistry in 1914, canceled his plans to attend the awards ceremony. He wished to have no contact with Haber or Planck, he wrote to one Swedish scientist, until they publicly repudiated the signatures they'd placed on the "Manifesto of the Ninety-three" in 1914.

There were no protests at the ceremony itself, which took place the following June, though many Allied diplomats and Nobel laureates found reasons not to attend. Just before the ceremony, Haber shaved the fugitive's beard he'd grown in Switzerland. In his acceptance speech, Haber spoke about the importance of nitrogen for agriculture, but didn't mention either gas weapons or the fact that ammonia mainly had been used, up to that point, as a tool for blowing things up.

But Haber was never a man who got much pleasure from looking toward the past. Almost always, he looked forward, impatiently searching for the next discovery, the next innovation. His speech ended on that note. "It may be that this solution is not the final one," he told his audience, noting that no scientist had yet managed either to understand or imitate the nitrogen fixation carried out by lowly bacteria in the soil. Unlocking that secret of nature might render every ammonia factory on earth obsolete. Haber's observation, in fact, remains true today, nearly a century later.

B y the time Haber received his Nobel medal, his mind already was stuck on another idea, another innovation that he hoped would rescue Germany. This time, the predicament was a financial one. The German mark was collapsing.

Inflation had begun even before the end of the war, as the German government began to print money to repay funds it had borrowed to finance the war. After the war ended, it accelerated. An item—in this case, one U.S. dollar—that cost 14 marks in July of 1919 cost nearly 200 marks by the beginning of 1922. "The mark falls constantly and the government is helpless!" Haber wrote to a friend.

Meanwhile, the war's victors demanded that Germany pay *their* war expenses, doing unto Germany as Germany had intended to do unto them. The Allies handed the Germans a bill for 132 billion gold marks—marks convertible into gold, and thus immune to inflation.

The crisis that this provoked was more political than financial, because the reparations came with no fixed timetable for payment. But the German government was caught between Allied force and popular rage. If Germany did not pay the hated reparations, France threatened to invade. If Germany did pay, the Germans themselves were likely to revolt, and not just at the polls. Nationalist groups assassinated two of Germany's leading moderate political leaders during these years.

In January 1923, France reacted to Germany's recalcitrance by marching troops into the Rhineland, determined to seize by force the coal that Germany wasn't turning over quickly enough. In response, the German government supported a campaign of passive resistance, essentially paying workers in the Rhineland to go on strike. It also paid for imported coal to make up for the coal no longer being mined in French-occupied areas. It financed all this with mountains of freshly printed German marks, and the rate of inflation went from rapid to mind-boggling.

One U.S. dollar, worth 200 marks at the beginning of 1922, could be exchanged for 18,000 marks at the beginning of the Rhineland crisis. Nine months later, it was worth 4 million marks, with no end in sight to the mark's downward spiral. In November, when the dollar reached a value of 4 *billion* marks, the government capitulated, gave up its resistance in the Rhineland, and established a new currency backed by solid assets.

Fortunes—including Haber's own, and the endowment that supported his institute—melted away to nothing in the furnace of inflation. Beginning in 1920, Haber labored to put his institute on a stable financial footing, transferring assets from increasingly worthless bonds into stocks, which were more likely to hold their value. He also tried to renegotiate his once-lucrative deal with the BASF, because the fixed payment of 1.5 pfennig per kilo of ammonia that the BASF manufactured (after 1919, .75 pfennig) was worth less by the day. Haber, however, had little bargaining power, and at first the BASF summarily rejected his request, leaving Haber bitter. Finally, in July of 1923, the BASF and Haber reached a settlement under which Haber received a large one-time payment.

Throughout this period, Haber worked obsessively on a secret project that promised to restore, as if by magic, Germany's financial health. The idea was born in 1920, when the foreign powers demanded reparations convertible into gold. Haber recalled something he'd read fifteen years earlier in a paper by the Swedish chemist Svante Arrhenius. Arrhenius had found tiny amounts of gold in seawater, and calculated that every ton of the ocean contained about 6 milligrams of the metal.

In the spring of 1920, Haber called a meeting of a few trusted colleagues at his institute. He swore them to secrecy, then announced his new goal: to find a way to extract tons of gold from the ocean.

The project—patriotic, practical, and enormously ambitious—was a typical Haber production. To the obvious question—if so much gold lay in the ocean, why hadn't anyone retrieved it already?—Haber had his own answer. Poor nations didn't have the ability to do it, and wealthy nations didn't have much interest in such a thing, since it would destroy the value of their existing gold reserves. Weimar Germany represented a new species of nation: technically advanced but poverty-stricken.

At first, prospects looked rosy. Tests on a few samples of seawater confirmed the estimates that Arrhenius had produced. The tests were crude and probably inexact, but there seemed to be plenty of room for error. Even if seawater actually contained only one-fifth that much gold, it would be enough to solve Germany's problems.

Haber decided to venture into the real world. He began making plans to collect water samples from the open ocean. He and several colleagues undertook two great voyages on commercial ocean liners, one to New York and the other to Argentina, collecting samples of ocean water along the way.

Those samples, however, delivered crushing news. They contained only a hundredth or a thousandth as much gold as Haber's scientists had previously observed in their laboratory.

It took three more years of painstaking research to clear up the mystery. The initial measurements that fed their early optimism, they learned, had been badly in error. Measuring extremely small concentrations of gold in water was far more difficult than they'd imagined. For one thing, the slightest contamination—from a laboratory worker who wiped his hands on the gold frames of his eyeglasses while conducting experiments, for instance—severely distorted the results. And there were traces of gold nearly everywhere, from the bottles in which seawater was stored to laboratory utensils.

Accurate measurements, carried out under extreme precautions to prevent contamination, drove the final stake through Haber's dreams. Arrhenius had been wrong. The average ton of seawater, rather than holding 5 or 6 milligrams of gold, contained only about 0.01 milligrams. In made no economic sense to extract such tiny quantities. In 1926, Haber shut down the gold campaign. That failure wounded Haber more deeply than he liked to admit.

Fritz Haber was not, by nature, a reflective man. He preferred to look forward, to solve whatever problem stood directly in his path, rather than ask why he was on that path in the first place. But events of the 1920s and his own increasing age forced Haber, more frequently as the years passed, to shift from striving to reflection.

There's a taste of it in two letters—one to Willstätter, the other to Charlotte and Hermann—that Haber sent from the deathbed of his father, Siegfried, in December of 1920. In both letters, but particularly in the one to his wife and son, Haber expresses admiration for his father, with whom he'd had such bitter clashes during his youth. In Haber's words there is wistfulness for his father's self-contained life—a life that Fritz Haber, in his ambition, had resolved to escape.

> Next door, in the large corner room, my father is gradually sleeping his way into death. . . . Yesterday when I came he still spoke with me, tired because of the morphine but with a clear spirit. He was serious and completely infused with the idea that he has lived out his life, and all his wishes are concentrated on the one hope that he "leave on the express train." . . . I am thinking about how much joy he gained from activities that were not for personal gain. . . . Many have this idealism. But, in addition, he had a confident self-restraint, so that in his public life nothing appeared tempting or worth doing that he would be unable to do or that exceeded his strength. . . . How strongly I feel his character now that he is leaving us.

In the summer of June 1924, Haber went home to Breslau once more to give the most reflective speech of his life. He'd been invited to mark the fiftieth anniversary of Breslau's Academic-Literary Association, a social club that had molded two generations of upwardly mobile middle-class children, many of them, like Haber, Jewish.

Haber's speech was not explicitly autobiographical; he did not describe his own ambitions and failures. Between the lines of his retelling of Germany's recent history, though, one hears self-criticism.

He recalled the naive complacency of his youth, when the laws and borders of Germany seemed as natural and unchanging as the tall medieval church that he passed every day. "It wasn't there for us to judge, but for us to adapt to, like day and night, like spring and winter, like everything towering and eternal." He spoke of unthinking German patriotism and officials who misused it as a "political weapon" to suppress dissent. "In those days it became fashionable among students to profess their loyalty to the fatherland on every occasion, as though Germany were in danger and had to be rescued once again through our devotion."

He looked back, this time critically, on the national celebration that accompanied Germany's economic boom. Power and wealth became society's goals, he said, "while the self-assurance of industry leaders rose heavenward along with the upward curve of iron production." But pride went before a great fall; the war destroyed both prosperity and illusions of national unity.

Haber looked out over the people in the audience and described the divisions among them, between those who longed for the return of an authoritarian state and those who defended democracy and the rights of workers. He took neither side in that argument, but ended his speech with a plea for tolerance, intellectual freedom, and democracy. He asked Germany's "warring brothers" to consider each other intellectual rivals to be convinced, rather than heretics to be destroyed. "Don't forget that only ignorance and old age give in to hate; they feel their weakness and their inability to persuade."

Fritz Haber, growing old himself, felt his own weakness. His letters to Richard Willstätter during the late 1920s are filled with complaints and anguish. Reading them, it's difficult to understand how Haber was able to accomplish anything at all during those years.

August 21, 1927

Dear Richard!

It's nighttime, and I'm afraid to sleep. The heart spasms, my latest achievement, only wake me when they've progressed to the point where I can't cut them off immediately with the alcoholic nitroglycerin solution. I don't know if it takes one minute or four before I've managed successfully to fiddle with the medicine dropper, put the drops on my tongue, and felt relief. But they're very bad minutes.

Nowhere else does Haber reveal similar depths of despair. Willstätter acted as Haber's father confessor, a stable rock on whom the more volatile Haber could lean. Haber was awed by Willstätter's calm and bottomless well of inner certainty; he sensed that he needed Willstätter more than Willstätter needed him.

"You asked how I'm doing. I'm suffering," Haber wrote in the spring of 1921. "Scientific writing is completely beyond my powers. . . . Medical examinations of all sorts have come to the conclusion that there's no organic cause of my ailments. So the doctors are inclined to diagnose a nervous disturbance of the pin-eal gland [a small hormone-secreting organ next to the brain]."

Whatever the cause, Haber felt that he was "living underneath a great boulder" most of the time—though not always. "Every once in a while it's as if a foreign power lifted the boulder so that the sun and air can temporarily reach me. In those hours I experi-ence life with gratitude."

Occasionally, suffering drove him toward uncharacteristic in-trospection, as in the first lines of a letter Haber composed on the final day of 1924.

> New Year's Eve, the day on which my mother died 56 years ago; the day that people celebrate because they've managed to get through another year. . . . I brood over the purpose of life. The only thing that's worthwhile for a man of my years and my nature is action; doing things; being useful. And I don't know where I can find a place that would make the necessary allowances for my damaged nerves and my diminished strength.

Others, though, saw a different Fritz Haber, charming, men-tally acute, and still the life of the party. Almost to the very end, Haber had the ability to gather himself and perform his accus-tomed public role, though he often paid the price later. Rudolf Stern remembered one such occasion, when he accompanied a weak and depressed Fritz Haber to a formal dinner in 1929. They ended up seated at a small table with Finance Minister Rudolf Hilferding and Hjalmar Schacht, president of the central bank. It was a socially awkward arrangement, for Schacht, who would later serve the Nazis loyally and lead Hitler's economics ministry, had just attacked Hilferding in print. But Haber "rose to the occasion like an old war horse which hears the drums of battle," Stern recalled. "For two hours, he treated us to a choice sampling of his famous anecdotes with such an irresistible charm that even Hilferding and Schacht could not help laughing and forgetting the dire facts of politics."

Haber was at his sparkling best during the regular colloquium that he hosted every second Monday in his institute, which became a fixture of Berlin's scientific life. These gatherings drew visitors not just from every corner of Berlin, but from across Germany and beyond. Haber made sure that the speakers didn't descend into the jargon of their own particular field. On one occasion, after a speaker named Weissenberg had gone on incomprehensibly for about five minutes, Haber suddenly rose from his seat and interrupted the lecture. "Mr. Weissenberg, there is a Thursday institution at which everyone is required to speak in such a manner that no one understands what's being said," he announced. Most people in the room knew that Haber was referring to meetings of the hallowed Prussian Academy of Sciences. "I urgently ask you not to transfer this peculiarity to Mondays." The room broke up in laughter, Dr. Weissenberg started over, and "we got to hear a very clear and informative lecture," recalled Lise Meitner.

Haber dominated these meetings not just because he hosted them, but because the format was a showcase for his particular kind of intelligence. He was a generalist, interested in absolutely everything, sharp-witted, and a bit of a showman. "He really could think aloud as he stood before his audience, in a way that aroused the wonder and admiration of all," said the British chemist John Coates. "One could never tell what would be his reaction to a new idea or to new experiments or theories; it was generally unexpected and always original and stimulating."

"He simply could react incredibly quickly," said James Franck, a colleague in Dahlem and a future Nobel laureate himself. "With him, there were none of those things you call 'stairway jokes'; Haber always knew what to say right away in the room, not later on the stairs. I've known even greater intellects—Albert Einstein and Niels Bohr. But I knew no one like Haber. This combination of—I'm tempted to say provocative—quickness in assessing a situation, along with good-heartedness and understanding, was quite remarkable."

Within his institute, Haber acted as patriarch and patron, wandering the halls and monitoring the work of young researchers. "Haber would come into my lab and question me: 'What have you done today, Dr. Alyea? Have you done so and so?'" remembered Hubert Alyea, who later taught at Princeton. "And then, no matter what I related, he would nod his head and say gravely, *'Aber! Aber! Aber!'* ['But! But! But!']" These chance meetings often turned into long lectures, fascinating in the morning but much less so at the end of the day, when they prevented researchers from leaving. Researchers who had plans for the evening were known to escape through ground-floor windows when they saw "the old one" wandering meditatively through the garden in the direction of their laboratory.

Haber's generosity was legendary. He seemed to regard anyone who passed through his institute as a personal dependent, and devoted himself to the task of helping them find their way professionally. As one friend put it, "A few more or less lame ducklings always followed in his wake, attaching themselves to him until he managed to find a place for them. His good nature was limitless when it came to such things." He also bought new clothes—"from head to foot"—every year for the children of the institute's gardener and its two mechanics, ten children in all.

The generosity wasn't entirely altruistic. Haber enjoyed passing out gifts and favors in part because it showed the world that he *could*. It demonstrated his own wealth and influence. As Lise Meitner once put it, Haber wanted to be "both your best friend and God at the same time."

Yet the scientists at the institute sensed Haber's loyalty and returned it. "He was the first person who, as an older man in a prominent position, took me completely seriously and gave me the feeling that I could talk to him in exactly the same way as to another assistant," said James Franck. When Lise Meitner wanted to hold a special seminar with Niels Bohr to which only younger scientists would be invited—a "bigwig-free" event, they called it—Haber immediately supported the idea. He did, however, ask Meitner if she would make an exception for himself and for Albert Einstein, which she did happily.

Despite his ailments, Haber never stopped building new institutions and propping up old ones. In addition to his duties at the Kaiser Wilhelm Society and the Prussian Academy of Sciences, Haber helped to create the Emergency Committee for German Science, which recruited research funding from private industry and government sources alike. He led efforts to heal the wartime division in European science, easing Germany's return to the International Research Council and the International Union of Pure and Applied Chemistry. He became fascinated by Japan and helped set up a Japan Institute in Berlin.

Haber often claimed to be weary of his many responsibilities. "A measure of leadership," he sighed, "is the number of requests that the person receives. The leader wants nothing from anyone, but others always want something from him." Occasionally, he played with the idea of giving up all of his positions outside the institute, or even leaving Berlin altogether. But he could never do it. His duties defined him; their burden gave his life meaning.

Willstätter, by contrast, was a free man. In the summer of 1924, his faculty colleagues at the University of Munich refused to appoint an obviously qualified Jewish candidate to a vacant professorship. Willstätter, thoroughly disgusted at rising anti-Semitism among the faculty, thought about it for a few hours, then resigned in protest.

Fritz Haber was dumbstruck. "The thunderclap will echo around the world," he wrote upon hearing the news. Ever the planner, he began thinking of Willstätter's career options. Perhaps rector of the university? Or perhaps a return to Berlin? Even as he wrote, Haber suspected that his proud friend would reject every proposal. "You have my support and my affection, no matter what you do, but I feel small and depressed when you don't give me a chance to do more than just to agree with you; to help organize things in some small way."

Despite appeals from faculty members and hundreds of students, Willstätter refused to reconsider. He also rejected prestigious offers elsewhere, feeling that accepting them would diminish his protest by turning it into a means of advancing his own career. Although he never entered his old laboratory again, he found a way to continue his scientific work. An assistant carried out the experiments at the university laboratories, and the two of them talked for an hour or more every day on the telephone.

Remarkably, Haber's friendship with Albert Einstein survived the war. Personal affinity triumphed over political differences. Together they faced the rising storm of anti-Semitism, some of it directed at Einstein himself. Their different instincts, however, shaped different responses. Haber insisted on demonstrating his loyalty to Germany; Einstein, who'd become an international celebrity, thumbed his nose at bigots and promoted Zionism.

On a few occasions, Haber tried to rein in Einstein's activism, mostly in vain. On March 9, 1921, a day after French troops occupied part of the Rhineland, inflaming German public opinion, he tried to persuade Einstein to postpone a trip to the United States and England. The visit was intended to raise money to build the Hebrew University of Jerusalem. Haber was sure that Germans would see it as fraternization with the enemy.

He tried to persuade Einstein that patriotic feelings were inevitable: "I know in my inmost being that as one grows older, the days come when certain things demonstrate their spiritual power: heritage and tradition and everything that gives order to life." By traveling to England and the United States at this moment, Haber continued, "you will proclaim to the whole world that you want to be nothing but a Swiss citizen who happens to live in Germany. . . . This is a time in which belonging to Germany brings with it a bit of martyrdom. Do you really want to demonstrate your inner alienation right now?"

This wasn't just a plea to come to the aid of Germany, Haber explained. He knew well enough that such a plea wouldn't carry much weight with Einstein. But like it or not, Einstein had become in the eyes of the world the foremost representative of German Jewry. His "ostentatious fraternization" with Germany's enemies would be seen by many Germans "as a sign of Jewish disloyalty." It would damage the reputation of German Jews for years to come.

Einstein replied the very same day with his usual ironic humor. "I'm not needed [in America] for my skills, of course, but because of my name," he wrote. "It holds great commercial promise among the rich tribesmen of Dollaria." The trip, he told Haber, simply couldn't be postponed. And anyway, considering the "perfidious and unloving" behavior of Germans toward Jews at the moment, he felt inclined to display more loyalty to his "tribe" than to Germany. "Dear Haber!" Einstein concluded. "An acquaintance recently called me a 'wild animal.' The wild animal likes you, and will try to visit you before his departure."

Einstein received job offers from all over the world during the 1920s, but he turned down each one. He didn't stay in Berlin because of any attachment to Germany, he told Haber, but rather because he couldn't bear to leave "my dear German friends, of whom you are one of the most outstanding and benevolent."

I n matters of love, Haber remained as hapless—or unlucky—as ever. In 1927, after ten years of strain and tumult, his marriage with Charlotte fell irreparably into pieces.

The break had been a long time coming. Yet Fritz, a man accustomed to command, never willingly shaped his life to accommodate the wishes of his wife—either one. And Charlotte had a strong will of her own.

Charlotte loved to travel. In 1922, she returned to Berlin from a trip and wrote this letter to her husband.

> Returning to our home in Dahlem was particularly hard after this trip. It's distressing to realize that every trip leaves me more alienated from this place. . . . There are dark shadows in this

house. There's no room for harmless jokes and fun. Light-heartedness can't just come from itself; it has to be thought-out and logical. The constant education, correcting both thought and action—it feels like pressure. As soon as this pressure is relieved, one's true nature explodes; the mask falls. That's the way it is, and this time I felt the difference between home and the outside world with special bitterness.

Charlotte fought to keep Fritz Haber at her side, but it was a losing battle. Too many things—work, friends, and long-established habits—pulled Haber in other directions.

Fritz's friends, judging by some of their later comments, never liked Charlotte very much. In his memories of Haber, Willstätter barely mentions Charlotte, or the two children she bore. (Charlotte and Fritz's son, Ludwig Fritz, was born in 1920.) Another friend called the marriage "unfortunate." There are hints of various reasons for these feelings: loyalty to Clara; distaste for Charlotte's boisterous personality; or a feeling that Charlotte interfered with something they considered more important, his science.

Charlotte also had a rival within her own home, her stepson Hermann, who was closer to her in age than Fritz was. Charlotte and Hermann never got along very well; as Hermann grew older and more independent, he waged an increasingly successful battle with his stepmother for primacy in his father's life.

The marriage, at least as Charlotte remembered it, bloomed anew in the fall of 1924, when Fritz and Charlotte left children and friends behind and embarked on a six-month cruise around the globe. The financial and political crises that had shaken Germany during the preceding four years were over for the time being. Haber himself, thanks to a lump-sum settlement with the

BASF the previous year, was on stable financial footing. Together they traveled through the United States to Japan, where they spent nearly two months, through the Far East, the Middle East, and the Mediterranean before returning to Berlin. Charlotte was entranced. The trip takes up more than a quarter of the pages in her autobiography.

Upon returning home, they received startling family news. Hermann, twenty-two years old and just finishing his university training, had fallen in love with Marga Stern, the sister of Haber's young friend and doctor Rudolf. In a sense, it was Hermann's first act of rebellion against his father. Fritz had directed every step of his son's life, practically ordering him to study chemistry instead of law, which Hermann preferred. A marriage so young, before Hermann had a chance to establish himself in his career, wasn't part of Fritz's plan. According to Charlotte, Rudolf Stern and Fritz also felt that both Hermann and Marga were personally unstable. Hermann had a weakness for drink; Marga was subject to depression. Yet the two would not be dissuaded. The marriage took place on January 14, 1926, after which the couple moved to the United States. Fritz arranged for Hermann to begin a series of short-term jobs in foreign chemical factories, just as Siegfried Haber had done for his son, Fritz, nearly forty years earlier.

Charlotte, rid of her rival, hoped for a cozy life with Fritz and their two children, but it was not to be. As always, there was work, travels, and Fritz's declining health, which took him—alone—to various sanatoriums for periods of rest. At the end of 1926, Fritz decided to relax in Monte Carlo with his friend Rudolf Stern for nearly a month, leaving her to pass the Christmas and New Year's holidays alone with the children. To make matters worse, Ludwig got very ill during that time.

Stern remembered the trip as an idyllic time, filled with walks and conversation. Haber's patron, the wealthy banker Leopold Koppel, now an old man but still "a delightful companion," also showed up. Charlotte, meanwhile, sat at home and cried. "I wasn't able to get over the bitterness created by this sad Christmas Eve, not for a long, long, time," she wrote.

The final break came the following summer, during a tumultuous week in a place that promised peace and quiet—a farmhouse in southern Germany that Haber owned, together with his lawyer and financial adviser, as a kind of rural retreat. The catalyst seems to have been the presence, once again, of Hermann. Charlotte, in her memoirs, recalled Fritz and Hermann playing chess while she sat apart, feeling "like the fifth wheel on a cart." She provoked an argument; Fritz erupted in a rage, and Charlotte left the room. The next day she fled to an art exhibition in Stuttgart, not far away. When she returned, Fritz had already departed, seeking refuge with Richard Willstätter in Munich. He had left behind a letter.

> I feared marriage with you, because our natures are completely different, and we couldn't grow together to the point where we could live a contented life together.... Your friends are not my friends; your inclinations are not mine. Even when we are together, we live for our own individual selves, and your attempts to change are as futile as mine. Let us call ten years enough. I can't do it anymore.

The feeling was mutual. Five months later, in December 1927, Charlotte and Fritz were legally divorced. Even though he confessed that separation from Charlotte was "necessary, if I was going to keep on living," it was evidently a source of great shame for Haber as well. For a time he avoided contact with colleagues. His sixtieth birthday arrived that same December, and German scientists filled a special issue of their leading journal, *Die Naturwissenschaften*, with essays about his life and work. In the courtyard of his institute they planted a linden tree in his honor. But Haber took little joy in the flood of cards and congratulations, and escaped them by traveling to Egypt with Hermann and Marga. There, he found little real diversion. "I don't want to stay here," he wrote to Willstätter on December 26. "I'm already restless, and the second self that the Egyptians call 'Ka,' which they say gives a person joy, flutters away from me for several hours every day toward my house and my institute."

The children, Eva and Ludwig, now spent most of their time with Charlotte, at her new home in central Berlin. Throughout the year, they visited their father's villa in Dahlem, and during the summer, they sometimes joined him at the farm in southern Germany. But they never felt particularly close to him; Fritz Haber seemed to them a feeble and distant figure. Haber certainly felt a keen sense of responsibility toward his younger children. It's uncertain, though, whether he loved them. Probably he did, but from a distance.

Fritz's half sister Else, whose husband had died a few years earlier, moved into his villa in Dahlem and managed the household. No one who knew Else had a bad word to say about her. She was, by all accounts, extraordinarily good and kind. Fritz and she were devoted to each other, and she remained at his side for the rest of his life.

Much of Fritz Haber's personal correspondence in the following years deals with money, for there was suddenly too little of it. In the divorce settlement, he'd agreed to maintain Charlotte and the children in the style to which they had been accustomed. He also wanted to support Hermann, who moved with his family from the United States to Czechoslovakia to France without finding a satisfactory job. But reckless investments in South American stocks claimed some of Haber's fortune, taxes ate up another chunk, and the world economic crisis of 1929 destroyed much of what was left.

Haber's distress over his worsening financial state fills his letters to Hermann during this period. It's not clear, however, whether he actually faced financial ruin or simply found losing so much of his previous wealth emotionally unbearable. Some portion of the fortune still remained; he had his salary as well, and the promise of a pension. When he traveled, he continued to do so in style, staying at the finest hotels of Paris, London, and Monte Carlo. On the other hand, Charlotte demanded that Haber fulfill at least the major portion of his financial commitments to her and the children, and Haber, in the midst of the worldwide economic collapse, struggled to do so.

In the midst of the economic crisis of the early 1930s, as Hitler's movement swelled in power, Haber's mood grew dark. On occasion, he even seemed to question technical progress, the enterprise to which he'd devoted his entire professional life. Early in 1932, he confessed that the previous half-century's technical innovations appeared to be merely "fire in the hands of small children."

One year later, Adolf Hitler was named Germany's chancellor, and Haber wrote the following lines to Richard Willstätter.

I battle with diminishing energy against my four enemies: sleep-lessness, the financial claims of my ex-wife, worry about the future, and the feeling that I've made serious mistakes in life.

Haber did not say what those mistakes had been.

ELEVEN

Dispossession

At a time when His Holiness, the infallible Pope of Christendom, is concluding a peace agreement, a Concordat, with the enemies of Christ, when the Protestants are establishing a "German church" and censoring the Bible, we descendents of the old Jews, the forefathers of European culture, are the only legitimate German representatives of that culture. Thanks to inscrutable wisdom, we are physically incapable of betraying it to the heathen civilization of poison gases, to the ammonia-breathing German war god.

—Joseph Roth, 1933

THE PORTLY, CIGAR-SMOKING PATRIARCH of Dahlem foresaw Germany's political catastrophe, but only its outlines. He never imagined that it could strip him so completely of his dearest possessions and turn his proudest accomplishments into ashes.

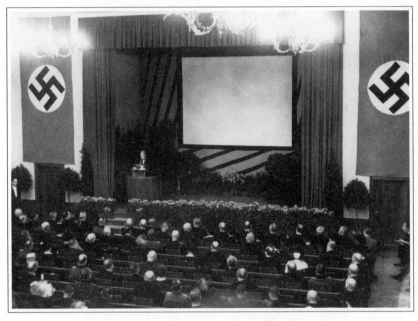

Max Planck speaking in the Harnack House in 1936.

As the world sank into economic depression and the number of unemployed workers in Germany soared past four million in 1931, Haber was disturbed to the point of sleeplessness over Germany's future. In a letter to the country's finance minister, he warned of a collapse "even worse than in 1918."

The crisis revived Haber's long-standing doubts about the viability and the fairness of capitalism. Those doubts had emerged first during World War I, when Haber saw war profiteering up close. On moral grounds, he disliked sermons that preached "the principle that every individual's selfishness leads to the greatest well-being of the whole." At the end of the war, he'd flirted with proposals for a kind of top-down socialism in which the state would own and control the country's major industries, but finally rejected the idea as unworkable.

Now he was ready to reconsider. "Back then, I felt that it was too early, and that such an idea demanded a new race of people. Well, the new race is here! It's filling the streets and pushing aside the established parties and perspectives. . . . The government will find this new race ungovernable if it does not either find a way out of the crisis *soon* with the current business leaders, or move into the new world."

Haber's letter to the minister of finance advocated extreme measures: If the government couldn't rejuvenate the economy with a flood of low-interest loans, it should renounce both capitalism and representative democracy, nationalize the country's factories, and put people back to work. "It may be foolishness," he wrote, but it was foolishness with the overwhelming force of public sentiment behind it.

———

A series of tottering Weimar coalitions couldn't manage such dramatic steps, especially against vigorous opposition from the wealthy and powerful. Haber's "new race" turned in droves to the brown-shirted nationalists led by Adolf Hitler. As Haber noted bitterly in a letter to his son, Hitler's National Socialist Party also managed to attract "a few millions" in campaign funding from sympathetic industrialists. In the 1932 election the Nazis became the single largest party in Germany, though they never received a majority of all votes cast in a national election.

The parliament lost its ability to govern, in part because many Germans, from Communists to conservative businessmen and especially National Socialists, didn't particularly want it to govern. They were ready to accept a führer, and on January 30, 1933, they got one. President Paul von Hindenburg, the aging World War I general, appointed Adolf Hitler Germany's chancellor.

There was now no stopping him, as Hitler set out to crush all opposition. In early February, Haber wrote that "I have to learn not to read the newspaper. It depresses me, because I see a view of life and the world taking over that is completely at odds with the thinking to which I'm accustomed."

Day by day, Hitler claimed new powers. On February 27, 1933, the Reichstag burned. A young Communist was arrested for setting the blaze, and the following day, the government suspended many civil liberties.

On March 23, German democracy completed its capitulation. The parliament handed Hitler's government absolute power. Even the party that Fritz Haber had supported, the liberal *Deutsche Staatspartei*, voted for the measure. Only the Social Democrats voted against it. Communist members of the Reichstag were prevented from voting because of unspecified security concerns.

It didn't take long for Haber to realize that he might become a target of Nazi repression. On April 1, the day on which Nazi leaders organized a nationwide boycott of Jewish businesses, Haber learned that the minister of justice in Prussia had demanded that all Jewish judges voluntarily request leaves of absence from their jobs. Any judges who refused, the minister announced, might be prevented from entering their courthouses by force.

Haber immediately sat down and wrote to Willstätter, pointing out that what happened to Jewish judges might also happen to Jewish scientists. He also recognized that the Nazis were employing their own definition of Jewishness. What counted now was ancestry alone; Haber's conversion was irrelevant. So he and Willstätter now sat in the same wildly rocking boat.

Less than a week later, Haber's forebodings became reality. The government unveiled a law ordering the removal within six months of all Jews from the German civil service, except for those Jews who'd been soldiers in World War I. The law covered every German university professor and nearly every scientist at the institutes of the Kaiser Wilhelm Society.

Haber, as a veteran of the war, could have stayed in office. At first, he was inclined to accept that privileged status. He could do more good by staying, he thought, aiding younger Jewish scientists in the search for new jobs. The thought of leaving may also have been too threatening. Haber's whole life was bound up in his institute. It had been his home, his work, and his community of friends for twenty-two years.

James Franck, who was in the same legal position as Haber, chose to resign his professorship at the University of Göttingen. In a hastily scrawled letter to Haber on April 15, Franck defended his decision, for he seems to have assumed that Haber would make the opposite choice. "I can't just get up in front of my students and act as though all this doesn't matter to me. And I also can't gnaw on the bone that the government tosses to Jewish war veterans. I honor and understand the position of those who want to hold out in their positions, but there also have to be people like me," Franck wrote. "So don't scold your James Franck, who loves you." Yet Haber, too, soon would find his situation intolerable.

L ike all civil servants, Haber received a questionnaire about the race of his immediate ancestors. Ever the punctual Prussian, he immediately sat down to fill it out. Asked for his race, he wrote "non-Aryan." But instead of dutifully listing the names of his parents and grandparents in the spaces provided, Haber scrawled these words diagonally across the page: "My parents and grandparents and both women to whom I've been married as well as their ancestors were all non-Aryan as defined by the law." With each scornful stroke of Haber's pen, he reclaimed the identity that he'd shed forty years earlier inside the city cathedral of Jena. Baptism had not, as he'd imagined it would, made him fully German. Nor had a lifelong devotion to nation and fatherland.

Apart from Einstein, who was traveling at the time and immediately declared that he wasn't coming back to Germany, Haber was the most prominent Jewish scientist at the Kaiser Wilhelm Society. As a result, Nazi officials singled out his institute for special repression. Instead of waiting until September, as the law required, they ordered the Kaiser Wilhelm Society immediately to "change the makeup" of Haber's institute by dismissing some Jewish scientists. (Roughly a quarter of the scientists at Haber's institute were of Jewish descent.) Leaders of the Kaiser Wilhelm

Society wanted Haber to follow those orders; they felt that lower-ranking Jewish employees had to be sacrificed in order to bolster the standing of the society's top scientists, including Haber himself.

Haber, in an act of human decency, resisted. He felt that less prominent scientists, being more vulnerable, deserved *more* protection, not less. Over the opposition of the Kaiser Wilhelm Society's general secretary, Haber chose to dismiss the institute's two most renowned scientists, Herbert Freundlich and Michael Polanyi. These were the scientists likely to have the greatest success finding positions abroad. (Polanyi already had an offer waiting from the University of Manchester, and Freundlich quickly landed a position at University College in London.)

Haber could take no more of this. A few days later, on April 30, Haber sent his own letter of resignation to the Prussian minister of culture.

My tradition requires that when choosing coworkers for a scientific post, I consider only the professional and personal characteristics of the applicant, without regard for their racial makeup. You cannot expect a man in his sixty-fifth year to change the thinking that has guided him for the past thirty-nine years of university life, and you will understand that the pride with which he served his German homeland all his life now dictates this request for retirement.

Haber's resignation was not immediately effective. He gave himself and his employers five months, until October 1, 1933, to find a successor. Max Planck, president of the Kaiser Wilhelm Society, tried to reverse Haber's decision. For several weeks Planck shuttled between various government offices, hoping to work out a compromise that would allow Haber to stay. He had no success with Culture Minister Bernhard Rust. "I'm finished with the Jew Haber," Rust is said to have replied. Planck then turned to Adolf Hitler himself. At a meeting on May 16, he tried to convince the führer that forcing valuable Jews to emigrate amounted to Germany's "self-mutilation." According to an account that Planck published in 1947, Hitler flew into such a rage that Planck could only leave the room.

The remainder of Haber's life is a chronicle of losses, recorded by Haber himself in tortured letters to friends and potential benefactors. Some of the losses were financial and physical: his villa and institute, his fortune, and his remaining reserves of strength. Equally shattering, though, was a kind of spiritual dispossession, the loss of his faith and identity.

Fritz Haber's character was a fabric woven from threads of loyalty and responsibility, and the strongest fibers in it were the black, red, and gold ones of German unity. The fabric now was torn in shreds. In the last year of his life, Haber clutched wildly at a few new or leftover threads—devotion to family, democracy, or even Zionism—but it was too late for him to weave himself a whole new identity.

The deepest tragedy in all this wasn't even his betrayal by the nation he'd served so loyally. It was the fact that his destruction was, in part, self-destruction. He'd followed, and sometimes led, the march into the pyre of World War I; he'd honored the "Spirit of 1914," the myth of mystic national unity that provided no space for dissent. (That myth was reborn in Nazi propaganda.) He'd helped feed the ravenous beast that was turning on him.

"I am bitter as never before, and the feeling that this is unbearable increases by the day," he wrote to Willstätter. "I was German to an extent that I feel fully only now, and I'm filled with incredible disgust at the fact that I can no longer work well enough to attempt a new position in a different country."

It took a few months for the full extent of Haber's dispossession to sink in. For another five months, he remained the institute's director. He busied himself with the fates of others, trying to help young Jewish scientists in his institute secure new positions abroad. When it came to his own future, however, Haber seemed incapable of making any decision. He wanted to leave Germany, but felt bound to it as well. He was too weak to take up full-time scientific work, but couldn't imagine retiring, either.

He wrote to a fellow chemist in London, asking about the possibility of a fellowship at an English university. Even while writing the note, Haber seemed to realize the difficulties involved in leaving Germany.

> I have never in my life courted any honors, and I am deeply ashamed to write this. But perhaps you will have some understanding for the feelings of an old man who was tied to his country for his whole life, but who now has the feeling that he has lost his homeland—a homeland that his ancestors and he himself served to the best of their ability.

Albert Einstein, already living abroad, observed Haber's suffering but felt little sympathy. Einstein's earlier disdain for all things German had hardened, under the influence of events, into fierce loathing. His letters to Haber display the satisfaction, bordering on gloating, of a man who'd finally won a long-running argument. "I can imagine your inner conflicts," he wrote to Haber in May 1933. "It is somewhat like having to abandon a theory on which you have worked for your whole life. It's not the same for me because I never believed in it in the least."

During the summer of 1933, Haber traveled to Holland, France, and England, looking for a scientific institution that could offer him an honorary position. He found none. The visit to London did, however, mark the birth of a remarkable, though short-lived, friendship with the Zionist leader Chaim Weizmann.

Haber had met Weizmann a few years earlier in Paris. (Haber's son Hermann brought the two men together; Hermann worked for Weizmann's brother-in-law.) At first, Weizmann regarded the man from Berlin as a political and ideological foe. He felt that "Haber was lacking in any Jewish self-respect. He had converted to Christianity and had pulled all his family with him along the road to apostasy." To Weizmann's surprise, however, he found Haber "extremely affable" and even interested in Weizmann's campaign to establish a Jewish homeland in Palestine.

So in July of 1933, when Haber called Weizmann on the telephone, Weizmann responded at once. He met Haber at London's Russell Hotel and found the once-proud chemist "broken, muddled, moving about in a mental and moral vacuum."

"I made a feeble attempt to comfort him, but the truth is that I could scarcely look him in the eyes," Weizmann wrote. "I was ashamed for myself, ashamed for this cruel world, which allowed such things to happen, and ashamed for the error in which he had lived and worked throughout his life."

Weizmann asked if Haber might consider moving to Palestine, to help establish scientific institutions there. Haber didn't respond to the suggestion immediately, but a seed was planted in his mind.

Over the following months, he became increasingly fascinated by Weizmann's audacious venture, but he also voiced doubts. The Jews of Europe, he wrote to Weizmann, still were bound to their separate nations—"I knew the battlefields on which French and English Jews shot down German Jews, where French and English socialists did the same to German Social Democrats. . . . The Russian Jews are ahead of us, because they suffered during the days when we in Germany achieved honor and prestige." The German Jews, Haber felt, still cared more about material success than the ideals of the Jewish prophets: "Anyone who expects peaceful citizens to act like Maccabees will surely end up in an asylum."

Back in Berlin, Haber felt the walls closing in. Already, officials of the Kaiser Wilhelm Society were being pressured to remove Haber from his villa. Haber still couldn't decide on his next step. "The old man is completely kaputt," wrote one of his young associates, the Jewish-Hungarian scientist Ladislaus Farkas. (Farkas would eventually find a new home in Palestine.)

Haber left Berlin again on August 3. He planned to visit Hermann in Paris, attend a scientific conference in Spain, and return home in twenty days. As his train pulled out of the Berlin station, Haber had no idea that he was saying farewell to Dahlem and Germany forever. For the next five months he would wander across Europe from one hotel room to the next, never finding rest, engaged in a never-ending negotiation with fate.

In Paris, a concrete offer arrived, the first one that Haber found truly attractive. In one of the bizarre twists that fill the final chapter of Haber's life, it had been arranged by his former foes, a trio of England's World War I gas warriors: Harold Hartley, Frederic Donnan of Imperial College in London, and Sir William

Pope of the University of Cambridge. They had convinced Cambridge to offer Haber an honorary position that demanded neither significant research nor teaching.

From his room in the Hotel George V, Haber composed his reply to Pope. He wrote in German, but took great care to use a Latin-style script instead of the traditional German style of handwriting that an Englishman would find difficult to decipher.

He told Pope that he was deeply grateful for the offer. It offered him everything he wanted: a chance to remáin active in science and possibly to acquire British citizenship.

> My most important goals in life are that I *not die as a German citizen* and that I not bequeath to my children and grandchildren the civil rights of second-class citizenship, as German law now demands that they accept and endure on account of their Jewish grandparents and great-grandparents. The second thing that's important to me is to spend my last years in a scientific community of people, with honor but without heavy duties.

And still he hesitated. The difficulty, according to Haber himself, was financial. If he settled in England without official permission, Germany could demand that he pay an "emigration tax" that would claim up to three-quarters of his remaining assets. And Haber couldn't bring himself to write off his fortune; he needed it not just for himself, but to support Charlotte and their two teenage children.

Much of Haber's wealth, however, was in Switzerland, so he could simply have ignored the tax. Yet Haber wasn't willing to do that, either. He could not tolerate the prospect of being branded in his homeland a tax evader. Despite everything that had happened, his identity and sense of worth remained irrationally bound to Germany. He struggled to cut himself loose, but succeeded only in part.

His mind was filled with fantasies of an honorable exit from Germany, perhaps through the intervention of people whose word would carry weight in Berlin. "Is it conceivable, from the English point of view, that this invitation could take the form of such a ceremonial honor that the British ambassador in Berlin might present it to the German Foreign Office? In this case, the Foreign Office might feel obligated, as an act of international courtesy, to support it," he wrote to Pope.

Haber also met Chaim Weizmann in Paris, and once again he felt drawn to the idea of supporting the Hebrew University of Jerusalem. There was, however, a problem: His friend Albert Einstein was feuding with Weizmann over the university's management (Einstein called the place a "pigsty") and Haber didn't want to take sides in their dispute. He sent Einstein a letter, urging him to meet Weizmann and patch things up between them. "I was never in my life so Jewish as now!" he wrote.

Immediately after posting the letter Haber traveled to a conference of the International Chemical Union in Santander, on the northern coast of Spain. Einstein's dismissive reply reached him there. "I'm especially glad . . . that your earlier love for the blond beast has cooled off a bit. Who would have thought that my dear Haber would appear before me as an advocate of the Jewish— and even Palestine's—cause!" Einstein refused Haber's appeal to meet with Weizmann; he had no desire to listen to more of the "old fox's" lies. He hoped Haber never would return to Germany, where the intellectual class "is made up of men who lie on their bellies before common criminals, and even sympathize with those

criminals to a certain extent. They couldn't disappoint me because I never had any sympathy or respect for them—apart from
a few fine persons (Planck 60% noble and [Max von] Laue
100%)."

Willstätter also was in Santander. The conference struck him,
in hindsight, as a "cruel dance of death." The assembled scientists
performed their customary professional repertoire with Europe
sliding toward an abyss and several of the scientists themselves,
like Haber, approaching their mortal end.

Haber labored mightily on his presentation, as he always did,
staying up until late at night to prepare it. Few, however, came to
hear it. Most of those present had come only out of courtesy, for
they didn't understand German anyway. "My poor friend took
medications beforehand, but could not escape one of his heart
spasms," Willstätter wrote. "He stopped and took some nitroglycerin, then, gasping and trembling, brought his speech to something resembling a conclusion."

The conference ended on August 18, and Haber's heart
problems were worsening rapidly. Any physical effort, whether
climbing stairs or even walking, was a great strain. On the eve
of his departure from Santander, Haber wrote to his thirteen-
year-old son, Ludwig Fritz, known as Lutz. Lutz and his sister, Eva, were both at a boarding school in southern Germany.
The letter bears sad testimony to an old man's dreams for his
children—children whom he knew, for most of their lives, from
a distance.

> I will in all probability move to England and live at the University
> of Cambridge. I would like you to go to an English boarding
> school. It would then be possible for you to become an English

citizen, which is the best opportunity that you could be granted in life. . . . As for Eva, I don't know if the most important thing for her is to live with her mother, or whether you feel it is most important for the two of you to be together. . . . In any case, it will be necessary in the coming years for both of you to learn to speak English and French as well as you now speak German. For your future lives, this is absolutely necessary in these times.

As he wrote this letter, Haber still intended to return to Berlin and secure "permission for an honorable separation" from Germany. As he wrote to Pope in Cambridge: "As soon as I have this permission, I will accept your generous invitation."

He left for Berlin, but at the last possible moment before reentering Germany, in the Swiss city of Basel, exhaustion forced him to interrupt his journey and rest. There, he heard again from Chaim Weizmann, who was vacationing not far away in the resort town of Zermatt at the foot of the Matterhorn. Weizmann invited Haber to join him. Haber shouldn't have accepted; his doctors had warned him to avoid high altitudes. Yet he threw caution to the wind and boarded a train for Zermatt, so irresistible was Weizmann's invitation.

Weizmann recalls Haber breaking into an "eloquent tirade" during dinner. "I was one of the mightiest men in Germany," he told the Zionist leader, but "at the end of my life I find myself a bankrupt. When I am gone and forgotten your work will stand, a shining monument, in the long history of our people."

Weizmann once again tried to persuade Haber to move to Palestine, praising its climate and the quality of the young researchers he would find there. "You will work in peace and honor. It will be a return home to you—your journey's end." And as

Weizmann recalls the conversation, Haber "accepted with enthusiasm."

Haber's enthusiasms, however, had lost their stamina. Like flickering candle flames, they were easily extinguished by circumstances or the shifting breezes of his own conflicting desires.

His doctors' warning about high altitudes proved correct. As Haber left Zermatt, homeward bound once again, he suffered a complete physical collapse and was taken off the train in the Swiss town of Brig. Haber thought it was an apoplectic fit; his doctor Rudolf Stern thought it probably was a heart attack. Once again he never made it across the German border.

Within a day or two he recovered well enough to walk, and traveled to a sanatorium in the Swiss town of Mammern. His sister Else Freyhan came from Berlin to look after him there.

Haber spent all of September and most of October at the sanatorium, gradually recovering his strength, occasionally making ambitious plans (a trip to Palestine by way of Zurich, Budapest, and Genoa, for instance) only to cancel them again.

He remained in Switzerland as the effective date of his resignation approached. Else Freyhan returned briefly to Berlin to clear furniture out of his villa. Haber gave most of it to friends at the institute. On October 1, 1933, he directed that the following note be posted on the institute's bulletin board. It amounted to an understated defense of Haber's own patriotism.

With these words I depart from the Kaiser Wilhelm Institute that was built by the Leopold Koppel Foundation . . . and which under my leadership for 22 years was dedicated to serving humanity in times of peace, and the fatherland in times of war. As far as I can evaluate the result, it was good and useful for science and for the nation's defense.

The strain of living in hotels and sanatoriums was wearing Haber down; he was desperate for a more settled existence. But where? For a man like Haber, living meant working, and there was no prospect of work in Germany. In his weakened state, he feared the long journey to Palestine. He settled on Cambridge as his goal, but he continued to worry about losing most of his fortune to Germany's predatory emigration tax. Like a mentally caged animal, his mind paced impatiently back and forth over the unsatisfactory legal and financial terrain available to him. Should he apply for exemption from the tax and risk rejection? Should he maintain a German residence for legal purposes while living as a visitor in England? If ordered to pay, should he do so or instead become a fugitive from German justice? The constant fretting produced "restlessness and nervousness that's made worse by circumstances that no one can change," wrote his son Hermann with mounting exasperation.

At the end of October, with his dilemma still unresolved, Haber joined Hermann and his family in Paris. A few days later, accompanied by his sister, he arrived in England and took up residence at the University Arms Hotel in Cambridge.

Sir William Pope did his best to make the homeless man feel at home. But exile is never easy, particularly for a man accustomed to control over his surroundings. "I fear that I didn't adequately realize what it means, at my age, to move into a foreign language and lifestyle," Haber wrote to Weizmann. "I miss the leadership functions that I built up over 40 years at home."

Haber still was preoccupied by efforts to escape the German

emigration tax, and hoped for salvation through the intervention of influential men. He climbed a "skyscraper of hopes" when the British ambassador in Berlin agreed to pay a visit to the German authorities. But the skyscraper came crashing down; the ambassador spoke to officials at the Prussian Ministry of Culture, which had no authority in such matters, and the visit produced nothing substantive.

From a lawyer in Berlin, Haber learned of another betrayal. Executives from I. G. Farben, the German chemical monopoly into which the BASF had merged, had promised to "break all relations" with Haber if he settled in England.

There weren't, in fact, any financial relations with I. G. Farben left to sever. Haber's emotional connections to the company, however, ran deep, all the more so because Carl Bosch, the young BASF engineer who'd supported his experiments in Karlsruhe thirty years earlier, now was I. G. Farben's chief executive.

Without Bosch's endorsement in 1908 and 1909, Haber might never have created ammonia; without Haber's success, Bosch might never have become one of the most powerful industrialists in Germany. Their fates were linked, and so were their names, for the synthesis of ammonia was known worldwide as the "Haber-Bosch process." The two men had drifted apart over the years, but Carl Bosch, to his credit, hadn't completely turned his back on Haber, even under Nazi rule. When Fritz Haber resigned his position in April of 1933, Bosch had been the only executive from the BASF to write a letter of sympathy and support. (At a meeting with Hitler, Bosch also had argued that persecution of Jews would harm German science. According to Bosch, Hitler replied, "Then we'll just work for a hundred years without physics and chemistry!") In December, Bosch had marked Haber's sixty-fifth birthday with a card.

From London, Haber wrote to Bosch with words of despair and pleading.

> You are now a man of decisive importance in the homeland that I've left behind in order to spend the latter days of my life in foreign parts. You offered me your aid, of your own accord, and I took your words seriously. I have never done anything, or said a single word, that would stamp me as an enemy of the men who rule Germany. . . . Won't you make it possible for me to live out these remaining years of pitifully diminished health and strength in peace and decency? I have children that are as dear to my heart as your children are to yours, and I would like to see them grow up in a country where their descent from me and my father doesn't erode their rights and their honor.

Carl Bosch's support for Haber had its limits. There's no record of any further attempt by Bosch to intervene with Nazi authorities on Haber's behalf, and Haber's letters went unanswered.

Bosch, in fact, had just signed a deal with the devil, a cost-plus contract with the Nazi government for large quantities of synthetic gasoline. The contract rescued I. G. Farben's enormous efforts to convert coal into gasoline on a large scale. The project had been losing large sums of money. The government contract made it, for the first time, profitable. Eventually, that gasoline would fuel Hitler's blitzkrieg.

Haber felt himself growing weaker and blamed the damp English climate. As so often, he sought to escape from his agonies through travel. He wrote to Rudolf Stern, asking his friend and doctor to accompany him to a sanatorium in Locarno, Switzerland. The two arranged to meet in Basel.

Haber wrote one last letter to Sir William Pope on January 25, 1934. This one, probably Haber's last letter, was laboriously composed in his awkward English.

> This is my last night in your country. I have taken more sleeping draughts than I should, without success. . . . I leave here with the full impression of your kindness shown to me and appreciated twice as much because in former life I did not win any right of your special regard. In coming to England it was my great hope to live long enough to become an English citizen and to leave this quality as an inheritance to my children. My present state of mind is not favorable for this hope. But I believe that an honorable end, if it overcomes me in your country, will be the right end of a life passed in the spirit of democracy which was dear to me from my youth. . . .
>
> Please believe me your friend, whatever may happen, and be convinced that English welfare means to me the welfare of the ideals which I believe in for the present and forever.
>
> Your friend, Fritz Haber.

On the afternoon of January 29, 1934, after two days of traveling by train and ship from England, Haber and his sister Else arrived in Basel exhausted. They walked slowly the half block from the train station to the Hotel Euler, a reputable establishment on the northwest side of a plaza in front of the station. There they met Hermann and Marga, as well as Rudolf Stern, who was frightened to see how much worse Haber looked. Even a few minutes of talking brought on severe chest pains.

Stern told Haber to lie down and examined the aging man. As usual, Stern encouraged his patient, telling him that rest in a milder climate would work wonders. And as usual, Haber "responded to this psychological treatment with amazing alacrity. Nothing would hold him in bed. He got up and came downstairs and—almost without interruption—discussed plans for the future."

Upon retiring for the night, however, Haber immediately called Stern to his bedside. He was gasping for breath, his heart failing once again, this time catastrophically. He lost consciousness and despite Stern's best efforts to revive him, died that night.

A few weeks before his death, Haber had written down a few wishes for his burial and given them to Hermann. He wanted his ashes to rest at Clara's side in Dahlem, he wrote, but not if the anti-Semitic movement in Germany made this impossible or inappropriate. In that case, it was up to Hermann to choose another final resting place and have Clara's ashes brought there.

Hermann chose to allow his father's ashes to rest in the city where his journey had happened to end. He found a burial plot in a cemetery on the northeastern outskirts of Basel, alongside Switzerland's border with Germany.

Richard Willstätter came from Munich to speak at a farewell ceremony at Haber's graveside on February 1. Hermann Haber, Else Freyhan, and Rudolf Stern were there as well.

If it seemed appropriate to add an epitaph to his gravestone, Haber had written to his son, it should be the following: "In war and peace, as long as it was granted him, a servant of his homeland." That homeland lay infuriatingly close, forever unreachable. Hermann chose not to include the words on his father's gravestone.

TWELVE

Requiem

Please forgive me, that I took so long to respond to your Christmas greetings. I don't dare bother you with letters that touch on emotions, for my feelings, as I watch what fate has done to you—they can't really be expressed in words.

—Richard Willstätter to Fritz Haber,

January 22, 1934

NEWS OF HABER'S DEATH reached Albert Einstein in the United States. He sat down to write a letter of condolence to Hermann and Marga Haber, and this time his words were sympathetic and devoid of all scorn.

Now almost all of my true friends are dead. One begins to feel like a fossil, not a living creature.

At the end, he was forced to experience all the bitterness of being abandoned by the people of his circle, a circle that mattered

The grave of Fritz Haber and Clara Immerwahr in Basel, Switzerland.

very much to him, even though he recognized its dubious acts of violence.

I remember a conversation with him; it must have been about three years ago after a meeting of the Academy of Sciences. He was quite incensed about the way he'd been shabbily treated during a vote, and to recover he went with me to the Schlosscafé on Unter den Linden. I said to him, a bit drolly, "Console yourself with me—your moral standing is truly enviable, and here I am happy and cheerful!" And this is what he said: "Yes, all of society never mattered to *you*." It was the tragedy of the German Jew; the tragedy of unrequited love.

Fritz Haber's death did not go entirely unnoted in Nazi-ruled Berlin. There was a speech in his honor at the Prussian Academy of Sciences, and the physicist Max von Laue wrote an admiring obituary for Germany's leading scientific journal, *Die Naturwissenschaften*. In his essay, Laue compared Haber to the legendary Greek figure Themistocles, who "went down in history not as the man banished from the court of the Persian king, but as the victor of Salamis." Similarly, wrote Laue, Haber would "go down in history . . . as the man who . . . won bread out of air and achieved a triumph in the service of his nation and all of humanity."

Several of Haber's friends went to Max Planck, president of the Kaiser Wilhelm Society, and proposed that the society plan a memorial service in Haber's honor. Planck, never a man for open confrontation, was caught between two opposing loyalties—to his friends and to his government. He hesitated, and did nothing for almost a year. Eventually, his better instincts won out. He began preparations for a memorial service to be held on January 29, 1935, the first anniversary of Haber's death.

On January 15, two weeks before the event, a letter from the German minister of education, Bernhard Rust, dropped "like a bomb" in the midst of Planck's preparations. Rust attacked the planned memorial service as "a challenge to the national socialist state" and issued an order prohibiting all civil servants, including university faculty, from attending the event.

Planck was stunned. Part of him, he wrote later, was inclined to cancel the event "for the sake of peace," yet he felt that doing so would destroy the organization's independence and self-respect. He wrote back to Rust, assuring Rust that he and his scientific organization intended no disrespect toward Germany's rulers: "The Kaiser Wilhelm Society for the Advancement of Science has demonstrated often enough, through word and deed, its positive attitude toward the current state and its allegiance to the Führer and his government."

He requested a meeting with Rust, and the Nazi official agreed. The two men met at Rust's villa at noon on Saturday, January 26.

They settled themselves in an enclosed patio, looking out over the gardens. Minister Rust complained about his physical ailments and suggested that there really wasn't much to discuss, since he'd already agreed not to block the scientific society's plans.

Rather than take the hint, Planck began to talk about Fritz Haber the man, the scientist, and the patriot. He drew a portrait of Haber that even a Nazi official might appreciate, as an instrument of Germany's march toward greatness, author of a scientific and industrial legacy that any German government should honor.

Planck noted in passing Haber's resignation in 1933, and suggested that it resulted from trivial disputes over personnel policies, rather than political dissent. But that was all in the past, Planck continued. Think of what endured. Consider the Leuna Works!

"Pocketing the enormous profits from the Leuna Works while refusing to give the customary recognition to its intellectual creator strikes me as disgraceful ingratitude," he told the Nazi leader.

This was apparently too much for Rust, who interrupted to say that Haber may have done a lot for Germany, but the National Socialist German Workers Party had done even more.

Planck assured Rust that he wouldn't think for a minute of disagreeing with that. There were many ways to serve the fatherland, and "one need not preclude the other." Then, perhaps sensing a conversational stalemate, Planck steered the conversation toward practical details, such as whether some of Haber's former colleagues would be granted permission to attend the Haber memorial. Rust promised to grant permission, but reneged on the promise a day later.

Three days later, under a cold gray winter sky, a crowd gathered in Dahlem. They stepped across cobblestones dusted with wet snow toward the Harnack House, clubhouse and meeting hall for the Kaiser Wilhelm Society.

On any normal day, the Harnack House was filled with men. This crowd, however, was oddly dominated by women.

They entered through massive wooden doors. Then, turning to the right, they ascended a winding staircase and entered the Goethe Hall, elegant and lined with tall windows.

Richard Willstätter, Fritz Haber's dearest friend, entered the room. The Nazi edict had no power over him because he was no longer an employee of the state; he'd resigned from the faculty of Munich's university ten years earlier in protest against anti-Semitism. Willstätter's beard was white and his thoughts were bitter. He scanned the crowd to see who had dared show up.

In Willstätter's mind, the many missing faces stood out more vividly than the ones before him. Albert Einstein and James Franck were in exile in the United States. Max von Laue was practically within earshot, ensconced unhappily in his office at the Kaiser Wilhelm Institute of Physics, just a short walk from the Harnack House. Willstätter recognized no one from Haber's own institute.

Haber's son Hermann was also missing. He'd confessed in a heartfelt letter to Max Planck that he could not in good conscience return to the Berlin neighborhood that his father had loved. "His mortal husk will only finally find rest," Hermann wrote, "when he can rest in Dahlem as a good German, and when his children and relatives can approach his grave as free Germans." Until that day arrived, "his memory, in my view, belongs to his family and his friends; not to the public, not to outside observers, and not to Dahlem."

Yet the hushed room was filled. A few stalwart souls who'd made no secret of their hostility toward the Nazi government had chosen to ignore the government's order. Lise Meitner, the discovery of nuclear fission still in her future, sat near the back. Her collaborator Otto Hahn was scheduled to give one of the speeches in Haber's honor.

Most notably, Carl Bosch was there. The powerful industrialist, chairman of the mighty chemical cartel I. G. Farben, was joined by a row of other executives from the chemical industry who'd come at his telegraphed request. Bosch had been unable or unwilling to help Haber during the great chemist's final months in exile, but he had not, at least, forgotten his debt to Fritz Haber.

The women who filled most of the room, meanwhile, were the wives of the scientists who'd been ordered to stay away. Their presence was a muted protest, evidence of both timidity and integrity. Haber's colleagues wished to pay their respects, yet did not dare to come themselves.

Planck moved to the podium. Outwardly, he appeared calm. Privately, he worried that Nazi thugs might yet appear to disrupt the meeting. As the last notes of a haunting string quartet by Franz Schubert faded away, Max Planck raised his arm.

"Heil Hitler!"

The memorial service proceeded as planned, unmarred by any disturbance. Max Planck closed his address with these words: "We reward loyalty with loyalty, and dedicate this hour to the honor of Fritz Haber: a great scholar, an honorable man, a fighter for Germany."

No German newspapers covered the event, having been ordered not to. Five thousand miles away, though, the *New York Times* alerted its readers to the fact that leading German scientists had gathered, despite official disapproval, to honor the man "who probably contributed more than any other single individual to Germany's power" during World War I.

At the moment of this memorial service, a seemingly innocuous piece of Haber's legacy could be found in cities of Europe as far east as Sofia, Bulgaria, in warehouses of the German Society for Insect Control. It was the insecticide called Zyklon B, a cyanide-based crystal that turned into vapor when exposed to air.

Haber had intended this poison to protect human life, not destroy it. During World War I, his institute had used an earlier version of this insecticide, hydrogen cyanide, to get rid of insect infestations in flour mills and granaries. But that gas was odorless; it gave no warning of its presence. People sometimes were exposed to it by accident and died. So in 1919, immediately after the war, Haber's institute developed Zyklon A, also derived from cyanide, but mixed with a foul-smelling gas that would warn people of the poison and drive them away. (The same principle is followed with natural gas used in homes today; it is mixed with a strong-smelling gas to warn people of leaks.)

In 1920, the inventors of Zyklon moved from Haber's institute to another research establishment nearby. (Haber maintained an informal relationship with these scientists and helped arrange funding for their laboratory.) In 1924, they introduced a new version of this insecticide that was easier to handle and named it Zyklon B.

For nearly twenty years, Zyklon B followed the path that its creators had envisioned, wiping out infestations of flour moths and other insect pests. Then, after Haber's death, came horrors: war; the rounding up of Jews and other human objects of hate into ghettos and fenced camps; an increasing crescendo of killing. Nazi leaders spoke the word "extermination," and the SS turned to the tools by which Haber had exterminated insects. They built human-scale gas chambers and demanded that chemists reformulate Zyklon B, taking out the foul-smelling substance that had served as a warning of its presence. The insecticide became a tool of death on a scale beyond all normal imagination. Members of Fritz Haber's extended family, children of his sisters and cousins, were hauled to those camps and killed, poisoned by the fruit of their famous relative's research.

THIRTEEN

The Heirs

*I must point out that this foreigner is the son of the famous
chemist Fritz Haber, recipient of the Nobel Prize for chemistry and inventor of war gas, who set up intensive production of asphyxiant gas in Germany during the last war.*

—A French official,

commenting on Hermann Haber's request

for French citizenship in 1935

FRITZ HABER'S ONLY DAUGHTER lives along a narrow
winding street overlooking the ancient city of Bath, in western England. Eva Lewis, born Eva Charlotte Haber, stoops
a bit when standing and walks cautiously, but her mind is clear
and sharp. Eva was Fritz Haber's rambunctious child, a daredevil
who never met a mountainside too steep for skiing. Her younger
brother Lutz (Ludwig Fritz), by contrast, buried himself in books.

She was eight years old and Lutz was five (and her father was
already fifty-nine and ailing) when Fritz and Charlotte were di-

Hermann Haber, Charlotte Nathan, and Fritz Haber on the steps of the Kaiser Wilhelm Memorial Church in Berlin, on the occasion of Fritz and Charlotte's wedding.

vorced. For the next seven years they saw Fritz Haber on weekends and holidays. "My image is of an old man—now mind you, when you're thirteen, anyone seems old—an old man who needed assisting going up the stairs, who must never be disturbed, but who was quite jolly when he decided to spend some time with *die kleinen Kinder*, the 'little children,' as we were called."

Eva Lewis keeps a photo album filled with mementos of a life that disappeared. There are photos of Fritz Haber's parents and uncles in Breslau, of Fritz Haber at his institute, of an elderly Fritz Haber with his teenage daughter at a holiday resort—probably the last time Eva saw him.

She escaped that life, and she seems to have escaped its emotional clutches. She describes her parents with detachment, humor, and little trace of any German accent.

Life went on, and it was a full life. Eva arrived in England with her mother and brother in 1936 at the age of seventeen. She studied agricultural science, married an Englishman, worked on a farm in Kenya during World War II, then settled in Bath.

Her brother Lutz, being male and carrying his father's name, felt the burden of Fritz Haber's legacy more keenly. Instead of walking in his father's footsteps or simply ignoring them, he circled back around and studied them with great care.

Lutz Haber became an economic historian and wrote two books on the rise of chemical industries in Europe and North America. One covered the nineteenth century, the other the years 1900–1930. There was a personal reason for ending in 1930, he confessed to his readers: "My father's work falls into the first thirty years of the century, and at the centre of his scientific achievement is the ammonia synthesis . . . which, in war as in peace, is essential to the modern state." But he devoted his attention only to that part of his father's work, and not Fritz Haber's campaign on behalf of chemical weaponry.

In 1968, when Lutz Haber already had established himself in his profession, he was invited to Karlsruhe for a ceremony honoring the memory of his father on the hundredth anniversary of his birth. To the embarrassment of the organizers, the ceremony was interrupted briefly by two young students who unfurled a banner with the following proclamation:

Celebration of a Murderer

Haber = Father of Gas Warfare

Lutz Haber's first reaction was simply to dismiss the protest as "lies or at least grossly exaggerated." But the passion of the protestors intrigued him, and an ember of curiosity about the military side of his father's life began to smolder in his mind.

A few years later, Lutz Haber looked up Harold Hartley, the British gas warrior who led the British investigation into Germany's use of gas weapons (and became a friend of Fritz Haber in the process). Week after week, Lutz Haber visited Hartley in his room at a nursing home, listening to Hartley's stories and recording many of them on tape. He began visiting unfamiliar sections of British archives, assembling a picture of his father's wartime activities. That labor grew into another book, *The Poisonous Cloud*, on the use of poison gas by all sides during World War I.

The book was a labor of filial devotion, but also of utmost detachment. As the younger Haber put it, with characteristic dryness, "The generation gap has its uses." He did not defend his father's work, but neither did he condemn it. Instead, he simply described it—the technology, tactics, and the effects of chemical warfare. He arrived at an intriguing conclusion. There was little convincing evidence that chemical weapons brought much military gain for either side. Perhaps, he suggested, the enormous efforts devoted to gas warfare during World War I grew out of unfounded faith in the power of poison.

Lutz Haber is still alive. He lives in Bath, close to his sister, but his mind, afflicted by Alzheimer's disease, lies beyond reach.

Eva Lewis has little contact with Fritz Haber's other descendants, the grandchildren and great-grandchildren of Haber's first wife, Clara Immerwahr, and their son, Hermann. It's a consequence in part of emigration to different countries, but also of the domestic civil war that raged during Haber's lifetime between Hermann and Charlotte Haber, stepson and stepmother.

"They were each fighting for Fritz Haber," says Eva Lewis. "First Charlotte had the upper hand, then Hermann." She pauses. "You know, it was Hermann who really suffered."

In the mid-1930s, after his father died, Hermann Haber tried to make a new home for himself, his wife, Marga, and their three young children in France. He had a job in Paris, but his future there was uncertain because he lacked French citizenship.

He applied to become a naturalized French citizen in 1935, and as he waited expectantly, French officials debated his case in private. One official advised against any quick approval of Hermann's request because his father, the "inventor of gas warfare . . . would normally figure in the list of those guilty of violations against human rights." Another official pointed out accusingly that Fritz Haber had been declared a "good German" at the memorial service in Berlin that Max Planck organized in 1935, implying that the loyal son of any good German must certainly be disloyal to France.

The obstinacy of this assertion boggles the mind. The memorial service had been an act of subdued protest against the Nazi government, and Nazi officials had prohibited state employees from attending it. Hermann himself had refused to attend, for he felt his father could not properly be honored in a country that had so completely rejected him. Yet French officials considered it evidence of his continued attachment to Germany. No matter which way he turned, Hermann Haber could not escape the shadow of his father.

The government summarily rejected his application in 1937. Hermann was informed that the French Ministry of War had vetoed it, but the precise reasons were never explained. Hermann wrote of mysterious "suspicions" against him, and his wife, Marga, wrote to a friend that Hermann's application had been turned down "because of his father." Hermann was crushed, she continued. "He believed that he could finally have gotten a real start here."

The rejection had fateful consequences. When World War II broke out, the French government classified Hermann Haber as an "enemy foreigner" and ordered him to an internment camp. Foreign residents were, however, allowed to fight with the French Foreign Legion, and Haber joined the Foreign Legion in order to escape internment. He was sent for a time to Algeria. When Germany invaded France, Marga Haber fled Paris with the children and found refuge in the Dordogne region in southern France. Hermann managed to join them there. They lived in constant danger, for under the terms of the armistice between Germany and France, any German or Jew living in France was to be turned over to German authorities.

Friends in the United States, especially Marga Stern's brother Rudolf Stern (Fritz Haber's friend and doctor) and Albert Einstein, mounted a campaign to save the Haber family and obtained exit visas for them. At the end of 1940, the family escaped from France, traveling first to Portugal and then to the Caribbean. The British interned them there for several months until Einstein again intervened. In June of 1941, Hermann, Marga, and their three daughters arrived in Hoboken, New Jersey.

Tragedy continued to stalk the family. Marga died of leukemia soon after the end of the war. Hermann, despondent, took his own life in 1946. The oldest of their three daughters committed suicide a short time later, the third member of this family to do so in three generations.

Fritz Stern, an eminent historian at Columbia University, became a guardian of Fritz Haber's legacy not because of genetic ties, but through friendship. Stern is Fritz Haber's godson, the son of Rudolf Stern, Haber's friend and doctor, who accompanied the aging and infirm chemist on his final trip.

Even after Haber's death, memories of the famous chemist and conversations about him filled the home where Fritz Stern grew up. Stern's parents were in awe of Haber's intellectual powers and grateful for the aid and encouragement he'd given them. "I grew up under the shadow of Uncle Fritz," recalls Stern. "I arrived in the United States in 1938, and Fritz wasn't the most fortunate name to have. But I couldn't give it up. I was always conscious of being Fritz Haber's godson."

Small framed photographs of Fritz Haber hang on the wall of the hallway leading to the front door of Stern's elegant New York apartment. One depicts the chemist in profile. In another he faces the camera, clothed in black coat and hat, eyes smiling, a cigar in his mouth.

Like Lutz Haber, Fritz Stern became a historian and explored the world of his parents, imperial Germany. He wrote about Gerson von Bleichröder, a Jewish banker who managed the fortunes of Germany's political leaders during the nineteenth century, and about Germany's "illiberalism"—its rejection of democracy, individual freedom, and tolerance.

The Nazi terror that drove some of the Stern family from their homeland and murdered others is not the focus of his writing, but it usually hovers just out of view, "an inescapable moral and historical conundrum," as Stern once described it. For forty years, Stern combed through the writings left behind by Germans of the nineteenth and early twentieth centuries, trying to understand their thinking, searching for the philosophical roots of Hitler's rise to power.

But he never forgot "Uncle Fritz." And when Fritz Stern accepted an invitation to speak in Jerusalem about "Einstein's Germany" in 1979, his research led inevitably back to Fritz Haber: "Gradually I came to realize that I could not abandon the subject."

Stern's first essay on Haber, delivered as a speech in 1986, sounded slightly defensive, as though the historian felt a need to protect his father's friend against unfair attacks. A longer and more reflective version, published in 1999, traced the parallel yet contrasting lives of Einstein and Haber. Stern described the two scientists as "fraternal opposites" in their politics and in their science: "Einstein exemplified the genius of theoretical conception, Haber that of immediate and practical achievement. Einstein was a solitary master, Haber an impresario of collective greatness." One senses in Stern's profile of the two men a profound respect for Einstein, yet greater sympathy for the passionate and tragic figure of Fritz Haber.

Fritz Haber left behind a technological inheritance as well, a bequest of machinery that shaped Germany and the world. Some of the oldest parts of the legacy, machines retrieved from the Leuna Works, stand in the lonely outskirts of Merseburg, a city a hundred miles southwest of Berlin, formerly part of East Germany. They are part of an open-air museum.

One machine still runs, though slowly. As a forty-ton flywheel stirs into motion, steel rods raise a plunger and suck air into a steel chamber that acts as the machine's lungs. When the plunger comes down again, the machine exhales, driving air out the other side.

In 1917, a pump exactly like this one drove the first clouds of hydrogen and nitrogen gas into reaction chambers of the Leuna Works and pushed ammonia gas out the other side. The ammonia was destined for Germany's hungry munitions factories. When the first railcar filled with ammonia left the Leuna Works, it carried the following message on its side, handwritten in chalk: "Death to the French."

The war ended, but the flywheels kept spinning, pumping out ammonia for democrats and Nazis alike. The Leuna Works grew into the largest chemical complex on earth, and German engineers looked for other valuable products of this new technology, the high-pressure reaction chamber. Caught in the grip of what one historian later called "technological momentum," they decided to convert coal into gasoline. It became one of the major research and development projects of the century.

The project turned out to be uneconomical; it was much more expensive to make gasoline from coal than from imported oil. But Nazi officials came to its aid. Hitler wanted Germany to be economically self-sufficient, and Nazi officials may already have been planning for wars to come. When war came, Leuna delivered gasoline fuel for the German blitzkrieg.

Allied bombers took aim at these pillars of German military power. Between May 12, 1944, and April 4, 1945, some eighteen thousand bombs rained down on the Leuna Works, eventually bringing production at the factory to a complete standstill. Several hundred workers were killed, many of them forced laborers and prisoners of war who weren't allowed into bomb shelters.

When Soviet troops arrived, they hauled away Leuna's most modern machinery to the Soviet Union as reparations. They didn't bother with the pump and steam-driven flywheel that now stand in Merseburg's museum, though. The ancient machines stayed in their accustomed places and resumed their familiar rhythm: one turn per second, day after day for the next forty years, delivering ammonia to East Germany, the new "workers' and peasants' state."

The Leuna Works and its neighboring factories in the towns of Schkopau and Bitterfeld acted as the foundation of an increasingly bizarre economy. Because they lacked dollars to pay for oil, Communist central planners maintained Hitler's policy of self-sufficiency. Everything that the rest of the world manufactured from oil—synthetic rubber, plastic, nylon, and gasoline—East Germany made from coal. The coal mines that supplied these factories devoured hundreds of square miles of East German territory, obliterating entire villages and leaving behind otherworldly landscapes, enormous craters filled with lakes as acid as vinegar.

I saw the Leuna Works for the first time in 1989. The complex extended as far as the eye could see across the horizon. The buildings were black with soot, smokestacks spewed white clouds, and no escape was possible from the smell of high-sulfur coal. Plant managers wearing pins of the Socialist Unity Party assured me that the Leuna Works had a bright future in capitalism. Others, after work hours in the privacy of their homes, described factories filled with equipment fit only for a museum.

As East German socialism collapsed, the flywheels and pumps of the Leuna Works slowed and stopped forever. In the mid-1990s, the old factories were demolished and carted off as waste. The technological momentum of the Haber-Bosch era finally was exhausted.

The first tank car of ammonia from the Leuna Works, ready to depart from Germany's munitions factories on April 28, 1917. Workers wrote "Death to the French" on its side.

FOURTEEN

Lessons Learned

Scientists have often been accused of providing new weapons for the mutual destruction of nations. . . . However, in the past, scientists could disclaim direct responsibility for the use to which mankind had put their disinterested discoveries. We feel compelled to take a more active stand now, because . . . nuclear power is fraught with infinitely greater dangers than were all the inventions of the past. . . .

—The Franck Report, 1945

HAD GERMAN POLITICS TAKEN a different turn, Fritz Haber might be considered a hero, and statues of him might now stand in prominent places. Instead, Haber became a tragic figure, trapped within the moral blinders of his time, unable to recognize the direction of history.

Haber's ambitions may now seem shallow and his loyalties misguided, yet there was nothing terribly unusual about them. Most people, now as then, swim with the current of public sentiment; most embrace technical progress; most support their homelands. The arrow of Haber's life traced an ordinary trajectory, yet flew with extraordinary force, further and more dramatically than most.

Haber could not foresee the ultimate consequences of the path he chose; perhaps it isn't fair to expect that he should have. But those consequences—the fateful prolongation of a senseless war, the invention of new methods for dealing out death—stand as a warning to all who follow.

During the grim decade that followed Haber's death, the reality of exile drove some of his closest friends to reconsider their previous allegiances—the loyalties that had also been Fritz Haber's.

While Hitler's armies ravaged Europe, Richard Willstätter wrote his memoirs from a refuge in Switzerland. As he struggled to make sense of the history he'd witnessed, Willstätter's thoughts turned to the earlier war in which he and Haber served. He searched for explanations for that disaster and eventually settled on society's inability to properly harness the power of technology, what Willstätter called "the machine." The machine, he wrote, "has sped up the pace of history. It has placed demands on the abilities of humans that humans are not equipped to satisfy." Willstätter wondered, and doubted, if humanity's ethical and moral leadership could keep pace with its technical achievements.

On the other side of the Atlantic, meanwhile, the man whom many once had considered Fritz Haber's scientific heir considered the morality of weapons of mass destruction. James Franck had been Haber's comrade-in-arms for two decades. He'd been among

Haber's scientifically trained "gas troops" in the trenches around Ypres in 1915. Later, he'd become one of the most prominent researchers at Haber's institute in Berlin. Franck had won his Nobel Prize in 1925. Many expected him eventually to take Haber's place as director of the Dahlem institute. Instead, Franck, who also was Jewish, resigned his university position a few days before Haber in 1933.

Exiled from Germany, Franck acquired a new position at the University of Chicago. There, he found himself recruited in 1942 to help develop another novel and powerful weapon—America's atom bomb.

Of the many famous scientists who joined the wartime Manhattan Project, Franck was one of the few who'd seen a war up close. He was also the only one who set a political price for his participation. Franck extracted a promise from Arthur Compton, one of the men in charge of the project: Before the atom bomb was used in combat, Franck would have a chance to present his views on its use to a responsible official of the U.S. government.

In the spring and summer of 1945, with the bomb almost ready to wreak its fearsome destruction, Franck held Compton to his promise. He had breakfast with Secretary of Commerce (later Vice President) Henry A. Wallace in Washington. And together with several other scientists from the atom bomb project, Franck authored a report that urged the government not to use its new weapon. That document, called the Franck Report, became the manifesto of a new movement.

The report argued that if the United States kept its nuclear weapons, and especially if it *used* them, other nations certainly would acquire the bomb as well. Down that road lay endless terror, for nuclear weapons made defense impossible; a world filled with nuclear weapons would be a world in which all were vulnerable.

Franck and his colleagues hammered at this point repeatedly: All the wonders of science could offer no protection and no escape from the dangers of nuclear war. Security was possible only through politics and diplomacy.

The United States, they argued, had to give up its ultimate weapon. A strong international authority, rather than individual nations, should control both nuclear weapons and the uranium necessary for building them. Only in this way could nuclear war be prevented.

It is hard to imagine a more complete departure from the mental world of Fritz Haber. For Haber, loyalty was a duty beyond question; Franck, his friend and protégé, saw duty in dissent. Haber had tried to win a war with ever-more-effective weapons; Franck saw the ultimate futility of a purely technological solution to the problem of national security. The émigré had torn off his mentor's moral blinders.

D ahlem, the place Haber never wanted to leave, still is a scientific community. After World War II, the Free University of Berlin was founded there. A visitor to Berlin can get there easily on Line 1 of the city's subway.

Straight ahead, after leaving the station, lies the Harnack House, where Fritz Haber once presided over his famous Monday Colloquium and where Max Planck led a memorial service in Haber's honor in 1935. Not far away, to the left, are the former villas of Richard Willstätter and Fritz Haber, and just beyond them lies Fritz Haber's beloved institute.

Crowds of students and teachers stream past these buildings every day, but their minds are preoccupied with struggles of the present, not the past. They pay little attention to historical markers that celebrate Dahlem's days of glory—or those that commemorate its shame, like the marker on a building where scientists studied internal organs of people murdered at Auschwitz.

Still, it's difficult to ignore the name of the largest research institute in the neighborhood, the Fritz Haber Institute. The institute, employing more than a hundred scientists, occupies the original building where Haber worked, along with several newer buildings nearby. It was renamed for Haber in 1952 at the insistence of the physicist Max von Laue, one of the last of Haber's friends who remained in Berlin.

"People often ask, 'Why don't you change the name?'" says Matthias Scheffler, one of the institute's five directors. Those who recognize Haber's name usually think of gas warfare, which isn't exactly the image that the institute wishes to cultivate. A high school in Berlin that once bore Haber's name did drop it a few years ago. It is now the Luise and Wilhelm Teske School, named for a shoemaker and his wife who hid Jews and other fugitives from the Nazis.

But Scheffler prefers to keep the name. It reminds every scientist at the institute that knowledge can be a tool for good and for evil, for creation and destruction. "They should understand that science has these two sides," he says.

I am especially glad . . . that your earlier love for the blond beast has cooled off a bit," Albert Einstein wrote to a dying and desperate Fritz Haber in 1933. Einstein was referring to an image in the writings of Friedrich Nietzsche. The iconoclastic German philosopher had written of a "splendid blond beast" that prowls in the spirits of all great nations, a kind of national id, untamed and bent on conquest.

Einstein's sarcasm was probably aimed at Haber's love for Germany. But there was another blond beast in Fritz Haber's life to which he was equally devoted, the spirit of technical innovation. Fritz Haber's deepest conviction, more central than his patriotism, was his faith in science and technical progress. Whatever the national dilemma—fertilizer shortages, British blockades on raw materials, trench warfare, or reparations payable in gold—he looked instinctively to technology for a solution. Only at the end of his life did Haber begin to question this faith, confessing to a friend that the great technical accomplishments of the previous half-century appeared increasingly "like fire in the hands of small children."

Haber's blond beast of progress still roams the globe, still restless, still claiming new territory of knowledge and power. Its conquests have been humanity's treasures—longer and more comfortable lives, freedom from sickness, ignorance, and misery. But the beast is a reckless and unpredictable creature, difficult to control. Left to its own devices, it is prone to wanton destruction.

Haber, a prophet of progress almost to the end, realized too late that the beast cannot be trusted with humanity's fate. It cannot be completely caged, but it needs to be watched and sometimes trained. The spirit of innovation cannot be stamped out—nor should it—but it can be directed and controlled by equally powerful human impulses of responsibility and love. Humanity cannot unlearn nuclear fission, for instance, but it can control the use of the world's uranium.

Nor is technology by itself usually the answer to humanity's most vexing dilemmas. In the midst of a senseless war, Haber answered every call from his nation's leaders, and delivered technology's best answers. But sometimes, it's the duty of an honest scientist to dash all hope that technology will rescue humanity from its folly. Sometimes, science cannot save.

Notes

PREFACE

xiv **could not survive in the absence of Fritz Haber's invention:** Smil, *Enriching the Earth*, pp. 155–60.

xiv **"But there was never a fruitful exchange of ideas":** Haber, *Fünf Vorträge*, p. 29.

xvi **"his generosity and reason always triumphed":** Meitner to Jaenicke, December 4, 1954, HC 1484 (Haber Collection, Archives for the History of the Max Planck Society, Dept. Va, Rep. 13).

ONE | YOUNG FRITZ

1 **"The influence of time and surroundings can't be denied":** Jaenicke notes from conversation with James Franck, April 16, 1958, HC 1449.

2 **"the extirpation of the German spirit for the benefit of the German empire":** Nietzsche, *Untimely Meditations*, p. 3.

3 **"lived from his memories":** Freyhan, "Erinnerungen an die Familie Haber und an Fritz Habers Jugendjahre," September 1953, HC 485.

4 **Siegfried, on the other hand, doted on his daughters:** Freyhan, "Erinnerungen," HC 485.

4 **"liked to air things out before everyone leaves":** Glücksman, "Fritz Habers Elternhaus," undated, HC 485.

4 **bubbled with fantastic ideas and creatively embroidered tales:** Freund, "Zu dem Stammbaum der Familie Haber," undated, HC 484.

4 **inherited nothing of his father's temperament:** Freund to Jaenicke, October 13, 1957, HC 485.

4 **"a characteristic that's been inherited by Fritz Haber and his generation":** Freyhan, "Erinnerungen," HC 485.

4 **acted as patriarch and domestic despot:** Stern, "Fritz Haber: Personal Recollections," *Yearbook,* Leo Baeck Institute, vol. 8 (1963), p. 72.

5 **"Our childhood and youth were illuminated by our brother's talents":** Freyhan, "Erinnerungen," HC 485.

5 **Fritz Haber's eyes filled with tears and his voice shook:** Stern, "Fritz Haber: Personal Recollections," p. 72.

6 **"I'm so disgusted with my entire life here that I could burst . . .":** Haber to Hamburger, May 25, 1885, HC 2305/1.

6 **"First, I'll go to university. . . .":** Haber to Hamburger, June 23, 1885, HC 2305/2.

6 **He dreamed of the "hours of genuine life" that he might experience there:** Ibid.

6 **allowed the budding chemist to use one of the rooms in her house:** Freyhan, "Erinnerungen," HC 485.

6 **as much as the total annual earnings of the best-paid mine workers of the time:** Szöllössi-Janze, *Fritz Haber, 1868–1934: Eine Biographie,* p. 37.

7 **convinced "that everything would get better":** Freund, "Stammbaum," HC 484.

8 **a mistake to overemphasize her brother's Jewishness:** Szöllössi-Janze, *Fritz Haber,* p. 683.

8 **"a form of communion," a passage into new life:** Klemperer, *Curriculum Vitae: Jugend um 1900,* vol. 1, pp. 44–45.

9 **Fifty years later, half of them did:** Landes, *The Unbound Prometheus,* p. 187.

10 **he confessed that he liked it that way:** Adams, *The Education of Henry Adams,* p. 83.

10 **"One great empire was ruled by one great emperor—Coal":** Ibid., pp. 414–15.

10 **"its limitless riches rose to the light of the working day":** Haber, *Aus Leben und Beruf,* p. 30.

11 **"this leap to hegemony, almost to monopoly, has no parallel":** Landes, *Unbound Prometheus,* p. 276.

12 **"we were naively bourgeois to our very bones":** Haber, *Aus Leben und Beruf,* pp. 29–30.

TWO | DIVERSIONS AND CONVERSION

15 **experiments that Haber found trivial and unchallenging:** Willstätter, *Aus meinem Leben,* p. 242.

15 **"have enticed me too often into rapids and eddies":** Haber to Hamburger, January 23, 1887, HC 2305/3. Also Haber to Hamburger, February 2, 1887, 2305/7. (Mistakenly filed with correspondence from 1891.)

15 **"man of proper responsibility in thought and action":** Jarausch, "The Universities," p. 187.

16 **nearly extinguished Haber's interest in chemistry:** Willstätter, *Aus meinem Leben,* p. 260.

16 **He complained of "nervousness":** Haber to Hamburger, December 21, 1887, HC 2305/3.

17 **only about a third actually chose this route:** Szöllössi-Janze, *Fritz Haber,* p. 45.

17 **pursued, unsuccessfully, the charming Julie Hamburger:** Ibid., p. 47.

17 **the "noisy desolation" of the firing range:** Ibid., p. 46.

18 **no Prussian Jew had ever become a reserve officer:** Ibid., p. 47.

19 **he called himself a "lousy product":** Haber to Hamburger, April 11, 1889, HC 2305/5.

19 **"You learn to be modest":** Haber to Hamburger, February 14, 1891, HC 2305/6. (Haber dated the letter 1890, but Szöllössi-Janze argues convincingly that this was a slip of the pen.)

21 **"never be satisfied with small assignments and small-minded objectives":** Willstätter, *Aus meinem Leben,* p. 260.

22 **"desert of sand, marsh, and fever. . . . There is nothing here, nothing at all":** Haber to Hamburger, September 25, 1891, 2305/7.

23 **his teacher was "authoritarian":** Willstätter, *Aus meinem Leben,* p. 260.

23 **"he'd never marry anyone else":** Rasch to unknown, February 1, 1955, HC 1493.

24 **bustling with visitors, year in and year out:** Radkau, *Das Zeitalter der Nervosität,* p. 114.

25 **"We fly through whole regions of the earth with the speed of the wind":** Ibid., p. 187.

26 **an "impossible alliance":** Stern, "Fritz Haber: Personal Recollections," p. 72.

26 **"a danger to the business":** Coates, "The Haber Memorial Lecture," p. 1642.

27 **Siegfried Haber was stuck with chemicals no one needed:** Goran, *Story of Fritz Haber,* pp. 14–15.

27 **"that's the reason I can't stand the Russians!":** Quoted in Plenz to Jaenicke, April 5, 1955, HC 1492.

27 **"I'm as comfortable as a drying fish":** Quoted in Szöllössi-Janze, p. 66.

28 **Switching one's religious identity seemed "ill-mannered":** Willstätter, *Aus meinem Leben,* p. 79.

29 **"seemed to him like a betrayal of one's ancestors":** Charlotte Haber, *Mein Leben mit Fritz Haber,* p. 84.

29 **was inspired by the Sermon on the Mount:** Ibid.

29 **"We felt 100% German":** Stern, "Fritz Haber: Personal Recollections," p. 88.

29 **essay by historian Theodor Mommsen:** A copy of the essay (and the essay by Treitschke that provoked it) can be found in Boehlich, *Der Berliner Antisemitismusstreit,* pp. 210–25.

30 **he molded the intellectual landscape of an entire generation of young Germans:** Jarausch, *Students, Society, and Politics in Imperial Germany,* pp. 208–12.

31 **"to destroy the barriers that divide them from their fellow German citizens":** Quoted in Boelich, *Der Berliner Antisemitismusstreit,* p. 224.

33 **"a beast of strength and sullen stupidity, the gentile?":** Fritz Stern, *Einstein's German World,* p. 76.

33 **fewer by far than in other regions:** Szöllössi-Janze, *Fritz Haber,* p. 97.

THREE | AMBITION

35 **"He went at a thing like a bull at a gate":** John Coates notes from conversation with W. H. Patterson, October 1934, HC 1425.

37 **"His name was barely known at the time":** Ostwald, *Lebenslinien: Eine Selbstbiographie,* vol. 1, p. 253.

38 **"and for that reason [lacks] a sure grasp of the subject":** Quoted in Stoltzenberg, *Fritz Haber,* p. 62.

38 **"studied every night until 2 a.m. until I got it":** Jaenicke notes from conversation with Max Meyer, November 9, 1958, HC 1483.

38 **"He taught himself":** Coates, "Haber Memorial Lecture," p. 1643.

39 **"what he needed most, the constant exercise and stimulus of discussion":** Ibid.

40 **"just four years earlier with only mediocre qualifications":** Szöllössi-Janze, *Fritz Haber,* p. 114.

40 **certainly have to move on to a different university:** Ibid., p. 115.

41 **publicly laid it bare for all to see:** Jaenicke notes from conversation with Max Meyer, November 9, 1958, HC 1483.

41 **"he knows even more. He's a know-it-all":** Quoted in Plenz to Jaenicke, April 5, 1955, HC 1492.

41 **effects of these retreats didn't often last long:** Szöllössi-Janze, *Fritz Haber*, p. 123.

FOUR | CLARA

43 **"I admire their colors and glitter, but I get no further":** Jaenicke notes from conversation with Max Meyer, November 9, 1958, HC 1483.

45 **"In any case, *auf Wiedersehen!*":** Haber to Richard Abegg and Clara Immerwahr, March 14, 1901, HC 923.

46 **an "intellectual aristocracy":** Jaenicke notes from conversation with Adelheid Noack, November 19, 1959, HC 301.

46 **She needed it, she wrote mysteriously, for "really quite simple reasons":** Immerwahr to Hamburger, April 29, 1891, HC 2305/15.

46 **"diligently but unsuccessfully" trying to forget about Clara:** Haber to unknown "Herr Professor," April 18, 1901, HC 1874.

47 **directly on Breslau's beautiful central square, facing City Hall:** Leitner, *Der Fall Clara Immerwahr*, pp. 26–27.

49 **"bitter feelings in my heart toward people who are dear to me":** Immerwahr to Abegg, March 31, 1900, HC 812.

49 **at the prospect of being "orphaned":** Immerwahr to Abegg, April 29, 1900, HC 812.

49 **"it's only because I'm so terribly happy!":** Immerwahr to Abegg, May 2, 1900, HC 812.

50 **"beautiful and holy duty within the shelter of the family":** Reprinted in Leitner, *Der Fall Clara Immerwahr*, p. 66.

50 **she "wasn't the right sort for marriage":** Charlotte Haber, *Mein Leben mit Fritz Haber*, p. 86.

51 **"prevailed upon to give it a try with me":** Haber to unknown, April 18, 1901, HC 1874.

51 **"would lie fallow and untouched":** Clara (Immerwahr) Haber to Abegg, April 23, 1909, HC 813.

51 **stricken on the way home by a heart attack and died:** Haber to unknown, April 18, 1901, HC 1874.

52 **"can't completely leave it behind, not even in my thoughts":** Clara Haber to Abegg, August 15, 1901, HC 812.

52 **"I think it was a physical thing":** Clara Haber to Abegg, October 10, 1901, HC 812.

53 **"I'd rather write ten dissertations than suffer this way":** Clara Haber to Abegg, February 2, 1902, HC 813.

53 **leaving Fritz in agony:** Clara Haber to Abegg, June 7, 1902, HC 924.

54 **enemy of true friendship and the "murderer of talent":** Jaenicke notes from conversation with Noack, November 19, 1959, HC 301.

FIVE | THE ENTHUSIAST

55 **"The seas inlaid with eloquent wires":** Walt Whitman, "Passage to India" (1868), in Malcolm Cowley, ed., *The Complete Poetry and Prose of Walt Whitman*, p. 361.

55 **a distant and possibly threatening land:** Fritz Haber, "Über Hochschulunterricht und elektrochemische Technik in den Vereinigten Staaten," pp. 291–406.

57 **"fear that we'll be overtaken in all areas":** Ibid., p. 302.

57 **"And the United States, a little bit":** Fritz Haber, "Vorträge," *Zeitschrift für Elektrochemie* 9 (1903), p. 894.

58 **"dwarf anything with which we are acquainted in the old continent":** Haber, "Über Hochschulunterricht," p. 291.

59 **operations laid out precisely for all the world to see:** Stoltzenberg, *Fritz Haber*, p. 75.

59 **"full of ambition to know and get on and miss nothing":** Coates, notes of conversation with W. H. Patterson, October 1934, HC 1425.

60 **a constant drive to save on labor costs and use more machines:** Haber, "Über Hochschulunterricht," p. 353.

60 **"The atmosphere of life is soaked in mechanical-technical ideas":** Ibid., p. 297.

60 **"America is for us among civilized countries the most distant":** "Vom Großbetrieb der Wissenschaft," in Harnack, *Aus Wissenschaft und Leben*, vol. 1, p. 19.

61 **"voluble in the strong metals and their infinite uses":** Quoted in Maier et al., *Inventing America: A History of the United States*, p. 564.

61 **"so that the students are there when the teacher arrives":** Quoted in Radkau, *Das Zeitalter der Nervosität*, p. 212.

61 **against the rules to take naps during working hours:** Ibid.

62 **"in the field of industrial production and economic expansion":** *Einweihung der Neubauten der Technischen Hochschule zu Karlsruhe*, May 17–19, 1899, p. 6, HC 1880.

62 **caught in the grip of a collective psychosis:** Radkau, *Das Zeitalter der Nervosität*, p. 216.

62 **"the way earlier generations dreamed of the hunt":** Ibid., p. 215.

63   **as harmless to a patient as ordinary light:** Ibid., p. 214.

63   **"The nationalists needed technical progress to strengthen the empire":** Ibid., p. 215.

63   **"he wouldn't even know that he was a father":** Quoted in Szöllösi-Janze, *Fritz Haber,* p. 142.

64   **"which Haber always found in working for the common good":** Coates, "Haber Memorial Lecture," p. 1649.

66   **"I don't want to be anyone's assistant":** Haber to Freund, June 11, 1905, HC 992a.

66   **"in view of his personal qualities":** Szöllössi-Janze, *Fritz Haber,* p. 151.

67   **"spin threads to all sides and cleverly tie them fast":** Wöhler to Jaenicke, May 1958, HC 1512.

67   **probably the result of Engler's quiet intervention:** Szöllösi-Janze, *Fritz Haber,* p. 152.

67   **frustrated chase through the city streets:** Witzeck to Jaenicke, October 7, 1960, HC 421.

68   **Haber showered them with questions and suggestions:** Witzeck to Jaenicke, October 7, 1960, HC 421.

68   **"he always moved in a kind of trot":** Ibid.

69   **Haber led the applause:** Ibid.

69   **"a great man, but friendly at the same time":** Jaenicke notes on conversation with Coates, September 26–28, 1959, HC 1438.

69   **"high-spirited and amusing":** Witzeck to Jaenicke, October 7, 1960, HS 421.

69   **"his home life was not too happy":** Fonda to Jaenicke, October 30, 1957, HC 123.

69   **"Haber was under continuous nagging from his wife":** Fonda to Jaenicke, February 12, 1958, HC 123.

70   **"He suffered a great deal from her pettiness":** Jahnke to Jaenicke, January 15, 1958, HC 215.

70   **"She sacrificed her profession for him":** Krassa to Jaenicke, November 2, 1957, HC 1470.

71   **"she really was good to her people":** Wöhler to Jaenicke, April 1958, HC 1512.

71   **"which family member might have caught what illness in which way!":** Jaenicke notes from conversation with Noack, November 19, 1959, HC 301.

71   **"she was a person with acute ethical standards, fanatically held":** Fonda to Jaenicke, February 12, 1958, HC 123.

71   **"Nothing could remain unspoken" or ambiguous:** Jaenicke notes from conversation with Noack, November 19, 1959, HC 301.

72 **"it would have been so interesting to be in England just now!":** Clara Haber to Abegg, August 25, 1907, HC 813.

SIX | FIXATION

73 **"achieved a triumph in the service of his nation and all of humanity":** Laue, "Fritz Haber," *Naturwissenschaften*, vol. 22 (1934), p. 97.

73 **to satisfy the whim of Germany's last emperor:** Hamecher, *Königin der See: Fünfmast-Vollschiff* Preussen, pp. 25–36.

75 **Chile's ports delivered more than a million tons of nitrate each year:** Smil, *Enriching the Earth*, p. 241. The figures quoted are metric tons.

77 **goal of agriculture was the packaging of nitrogen for human consumption:** Ibid., p. 8.

77 **"impudent humbug" produced by a "set of swindlers":** Quoted in ibid., p. 12.

80 **he proposed to introduce a new topic, an unfamiliar one "of urgent importance":** The speech is reprinted in Crookes, *The Wheat Problem*, pp. 1–41.

82 **"races to whom wheaten bread is not the staff of life":** Ibid., p. 38.

82 **"technology will make possible the artificial preparation of foodstuffs":** *Einweihung der Neubauten der Technischen Hochschule zu Karlsruhe*, May 17–19, 1889, HC 1880, p. 6.

82 **thought for one breathtaking moment that he'd solved the problem:** Stoltzenberg, *Fritz Haber*, pp. 140–43, and Ostwald, *Lebenslinien*, Part II, pp. 284–87.

84 **the Margulies brothers offered to pay him well:** Haber to BASF, February 20, 1908, HC 2069.

87 **"Now the situation is much less favorable":** "XIV. Hauptversammlung der Deutschen Bunsen-Gesellschaft für angewandte physikalische Chemie," *Zeitschrift für Elektrochemie*, vol. 13 (1907), pp. 523–24.

89 **place their talents at the service of foreign companies:** Haber to BASF, February 20, 1908, HC 2069.

89 **"not out of confidence in the matter itself":** Haber to BASF, October 16, 1914, HC 2115.

90 **licensing any future ammonia-related technology to the Austrian firm:** Haber to BASF, February 19, 1908, HC 2069.

90 **"he naturally can't be had on the cheap":** Engler to BASF, February 16, 1908, HC 2069.

90 **"equal to 10 percent of the company's net profits from his discoveries":** Szöllössi-Janze, *Fritz Haber*, pp. 173–74.

91 **render the entire nitrate trade obsolete:** Haber to BASF, May 20, 1908, HC 2072.

92 **ran through a narrow tube into a container below:** Haber to BASF, March 23, 1909, HC 2079.

93 **"It was fantastic":** Jaenicke notes from conversation with Staudinger, July 27, 1958, HC 1505.

93 **most of which lay in the hands of the Auergesellschaft:** Haber to BASF, March 23, 1909, HC 2079.

93 **that would blow up any commercial-scale reaction chamber:** Le Rossignol, "Zur Geschichte der Herstellung des synthetischen Ammoniaks," *Die Naturwissenschaften*, vol. 16 (1928), p. 1071.

93 **"I know exactly what the steel industry can do. It's worth risking":** Holdermann, *Im Banne der Chemie*, p. 69.

94 **Mittasch was "deeply impressed and completely convinced":** Coates, "Haber Memorial Lecture," p. 1653.

95 **"we could only walk in a straight line by following the streetcar tracks":** Jaenicke notes from conversation with Coates, September 26–28, 1959, HC 1438.

SEVEN | MYTHS AND MIRACLES

102 **from five million tons a year in 1960 to fourteen million tons in 1970:** Smil, *Enriching the Earth*, p. 243.

102 **A typical acre of English pasture now receives around 160 pounds of nitrogen annually:** Ibid., p. 148.

103 **"chemical fertilizer is the fuel that has powered its forward thrust":** The lecture can be found on the Nobel website: http://www.nobel.se/.

103 **four times greater than would be possible if they relied solely on the natural recycling:** Smil, *Enriching the Earth*, pp. 152–54.

103 **to feed all six billion of us our accustomed diet:** See ibid., pp. 155–60.

104 **"the company does not wish its identity known at this time":** E. J. Mitchell, "Trip Report: Peking, China," *Chemical Engineering*, July 9, 1973, pp. 92–98.

105 **"They simply could not produce any more!":** Interview with Vaclav Smil, June 2002. Excerpts broadcast on *Morning Edition*, National Public Radio, July 10, 2002.

105 **"eating meat and fish no more than a few times a year":** Smil, *Enriching the Earth*, p. 169.

106 **each acre of farmland in China's coastal provinces:** Ibid., p. 147.

106 **Yields of wheat have tripled over the same period:** Statistics drawn from the database of the Food and Agriculture Organization at http://apps.fao.org.

107 **In Africa, however, they are now rising:** Rosegrant et al., *Global Food Projections to 2020*, pp. 36–44.

108 **average depletion comes to more than sixty pounds per acre:** Henao and Baanante, "Nutrient Depletion in the Agricultural Soils of Africa," *2020 Brief No. 62*, p. 1.

110 **only four atoms enter a human mouth:** Galloway and Cowling, "Reactive Nitrogen and the World: 200 Years of Change," *Ambio*, vol. 31, no. 2 (March 2002), pp. 64–71.

110 **A million and a half tons of nitrogen each year float back down the Mississippi:** Goolsby et al., "Nitrogen Input to the Gulf of Mexico," *Journal of Environmental Quality*, vol. 30 (2001), pp. 329–36.

110 **at a rate three times higher than was the case a century ago:** Smil, *Enriching the Earth*, p. 180.

111 **more than most African farmers could dream of buying:** Ibid., p. 194.

111 **supporting a less rich and complex web of life:** Ibid., p. 195.

114 **"Consider the other side!":** Clara Haber to Abegg, April 23, 1909, HC 813.

EIGHT | EMPIRE CALLS

117 **"it will be not just equal to them; it will win new peaceful victories!":** Harnack, untitled presentation to Kaiser Wilhelm II, November 21, 1909, HC 1603.

117 **"unless we utilize science to the full for military and industrial purposes":** Hale to President Woodrow Wilson, March 26, 1918, Wilson manuscripts, Library of Congress, File VI, Case 206.

118 **built on steel, electrical equipment, chemicals, and banking:** Minutes of a meeting to plan the establishment of a society for the promotion of science, May 6, 1910, HC 1625.

119 **"vital interest" of the German nation:** Harnack, untitled presentation to Kaiser Wilhelm II, November 21, 1909, HC 1603.

121 **license those patents and profit from them:** Szöllössi-Janze, *Fritz Haber*, p. 218.

122 **refusal would have wrecked well-laid imperial plans:** Haber to Böhm, September 19, 1910, HC 941.

122 **he saw the emperor nod his head in agreement:** Johnson, *The Kaiser's Chemists*, p. 126.

122 **"without yet knowing what exactly will come out of it":** Quoted in Szöllössi-Janze, *Fritz Haber*, p. 225.

122 **The eagle was carrying him toward an unknown fate:** The image is reproduced in Stoltzenberg, *Fritz Haber*, p. 271.

123 **"from great researcher to great German"**: Willstätter, *Aus meinem Leben*, p. 263.

124 **"his fingers were itching to do it"**: Jaenicke notes from conversation with James Franck, April 16, 1958, HC 1449.

125 **Germans remained oblivious to their country's growing political isolation**: Mommsen, *Bürgerstolz und Weltmachtstreben*, p. 337.

125 **"wouldn't let the good German sword rust"**: Ibid., p. 468.

125 **voiced full-throated support for the chancellor's words**: Jarausch, *The Enigmatic Chancellor*, p. 125.

126 **"It is false that in Germany the nation is peaceful"**: Ibid., p. 124.

126 **"The twilight of the gods of the bourgeois world is in prospect"**: Quoted in Mommsen, "The Topos of Inevitable War in Germany in the Decade before 1914," in Volker Berghahn and Martin Kitchen, eds., *Germany in the Age of Total War*, p. 31.

126 **Other German industrialists also counseled patience**: Mommsen, *Imperial Germany, 1867–1918*, pp. 91–94.

127 **was dismayed by Stinnes's "delusions"**: Feldman, "Hugo Stinnes and the Prospect of War," in Manfred Boemeke, Roger Chickering, and Stig Förster, eds., *Anticipating Total War*, pp. 92–93.

127 **34 percent of all votes in the 1912 elections**: Mommsen, *Imperial Germany, 1867–1918*, pp. 91–94, 155.

127 **war "had an almost magical attractive power"**: Radkau, *Das Zeitalter der Nervosität*, pp. 461, 501.

127 **society's "lethargy and emasculation"**: Mommsen, "Topos of Inevitable War," p. 26.

127 **"strengthening bath of steel"**: Radkau, *Das Zeitalter der Nervosität*, p. 441.

128 **"idealistic belief that the German people need a war"**: Quoted in Fritz Stern, *The Failure of Illiberalism*, p. 154.

128 **"Russia is systematically preparing for war against us"**: Quoted in Mommsen, "Topos of Inevitable War," p. 38.

128 **"what force conquers, force alone will never hold"**: Quoted in Jarausch, *The Enigmatic Chancellor*, p. 145.

129 **friendly words to the man from Vienna**: Stern, "Fritz Haber: Personal Recollections," p. 73.

130 **"such an ideal [of academic freedom] simply cannot be realized"**: Reichstag transcript of April 25, 1910, HC 1625.

131 **named Dahlem's streets for pioneering chemists and physicists**: Willstätter, *Aus meinem Leben*, p. 204.

132 **"observe, close up, your research taking shape"**: Haber to Willstätter, 1911, in Petra Werner and Angelika Irmscher, eds., *Fritz Haber: Briefe an Richard Willstätter, 1910–1934*, pp. 49–50.

132   **"In your case I do sense this rare and fortunate event"**: Haber to Willstätter, July 31, 1911, in Werner and Irmscher, *Briefe*, p. 47.

132   **"was that I felt myself a German"**: Willstätter, *Aus meinem Leben*, p. 200.

133   **"the old-time venerable type of great Jewish Rabbi"**: Weizmann, *Trial and Error*, p. 350.

133   **called Willstätter an out-and-out nationalist**: Hevesy to Jaenicke, July 27, 1958, HC 1505.

133   **"severe bite wounds to our friendly though loud Bobbi"**: Willstätter, *Aus meinem Leben*, p. 204.

134   **"It's so much easier to say *'Du Esel'* ['You ass!']"**: Charlotte Haber, *Mein Leben mit Fritz Haber*, pp. 109–110.

135   **cover half of Einstein's salary for twelve years**: Szöllössi-Janze, *Fritz Haber*, p. 253.

135   **"wanted to get hold of a rare postage stamp"**: Stern, *Einstein's German World*, p. 112.

136   **"(raw in speech, movement, voice, feeling)"**: Ibid., p. 64.

137   **Einstein spent that night in Haber's home**: Ibid., p. 65.

139   **he would return immediately to Berlin**: Haber to Minister of Intellectual and Educational Matters, July 28, 1914, HC 1668.

NINE | "THE GREATEST PERIOD OF HIS LIFE"

141   **"I was one of the mightiest men in Germany"**: Quoted in Weizmann, *Trial and Error*, p. 354.

141   **"is systematic utilization of the scientific expert"**: Dewey, "What Are We Fighting For?" (1918), in *The Middle Works*, vol. 11, p. 99. My thanks to Christoph Strupp of the German Historical Institute for this reference.

142   **"God will help the German sword to victory"**: Quoted in Verhey, *The Spirit of 1914*, pp. 65–66.

142   **"brotherhood, belief, and certainty of victory"**: Quoted in ibid., p. 5.

142   **"Kaiser and *Volk*, government and citizens—all were one"**: Quoted in ibid., pp. 66–67.

143   **quiet and fearful talk, and a "deadening seriousness"**: Ibid., pp. 68–71.

143   **"the highest reality: The experience of belonging together"**: Quoted in ibid., p. 5.

144   **civilians would so brazenly interfere in military matters**: Holdermann, *Im Banne der Chemie*, pp. 136–37.

146   **convert ammonia into nitrate on a large scale**: Szöllössi-Janze, *Fritz Haber*, p. 284.

147 **"a hundred tons of nitric acid per day?"**: Holdermann, *Im Banne der Chemie*, pp. 136–37.

150 **the Germanic martial "Hero"**: Mommsen, *Imperial Germany, 1867–1918*, pp. 209–10.

150 **"certain in the knowledge that they are the people of God"**: Quoted in ibid., p. 212.

150 **"a deep-seated reluctance to look facts in the face"**: Ibid., p. 214.

150 **find the contacts he made there extremely useful**: Szöllössi-Janze, *Fritz Haber*, p. 307.

151 **he called the war "madness"**: Stern, *Einstein's German World*, p. 115.

151 **"began to forge a more powerful ax"**: Ibid., p. 118.

151 **"the diffusion of asphyxiating or deleterious gases"**: L. F. Haber, *The Poisonous Cloud*, p. 18.

152 **they'd been subjected to the world's first gas attack**: Ibid., pp. 24–25.

152 **in combat during the spring of 1915**: Ibid., pp. 23–24.

153 **his mind leaping quickly from laboratory experiments to factory-scale feasibility**: Jaenicke notes from conversation with Franck, April 16, 1958, HC 1449.

154 **"true vocation in executing military organizational tasks"**: Quoted in Stern, "Fritz Haber: Personal Recollections," p. 77.

154 **"an extremely energetic organizer, determined, and possibly unscrupulous"**: L. F. Haber, *The Poisonous Cloud*, p. 27.

155 **Clara proved to be "very calm and courageous"**: Lütge to Jaenicke, January 6, 1958, HC 1479.

155 **were observed weeping uncontrollably**: Sime, *Lise Meitner: A Life in Physics*, p. 58.

156 **"it would only hurt us"**: Quoted in Martinetz, *Der Gaskrieg 1914–18*, p. 18.

156 **trenches that surrounded the Belgian city of Ypres**: L. F. Haber, *The Poisonous Cloud*, p. 27.

157 **"Then he proceeded with a lecture about clouds of chlorine"**: Hahn to Jaenicke, January 12, 1955, HC 1453.

157 **"innumerable human lives would be saved if the war could be ended"**: Ibid.

157 **"*Geheimrat* Haber was accompanied by his wife"**: Lummitsch, "Meine Erinnerungen an Geheimrat Professor Dr. Haber," July–August 1955, HC 1480.

159 **pounded them with cannon shells**: L. F. Haber, *The Poisonous Cloud*, p. 31.

159 **Franck sat calmly at the bottom of the crater:** Lummitsch, "Erinnerungen," HC 1480.

159 **blowing inland from the coast:** Ibid.

160 **"*Geheimrat* Haber, though, looked very unhappy":** Ibid.

160 **"The higher civilization rises, the viler man becomes":** Quoted in Trumpener, "The Road to Ypres," *Journal of Modern History*, vol. 47 (September 1975), p. 473.

161 **"convinced of the weapon's terrible effectiveness":** L. F. Haber, *The Poisonous Cloud*, p. 31. Also Lummitsch, "Erinnerungen," HC 1480.

161 **deadly as the opponents of gas warfare claimed:** Haber, *Fünf Vorträge*, p. 88.

161 **"much too complicated for me to get up again":** Jaenicke notes from conversation with James Franck, April 16, 1958, HC 1449.

161 **never quite understood the implications:** Dieter Martinetz, *Der Gaskrieg 1914–18*, pp. 25–26.

162 **"this passed off fairly soon":** Macdonald, *1915: The Death of Innocence*, p. 194.

163 **"You could recognize the famous Cloth Hall":** Lummitsch, "Erinnerungen," HC 1480.

164 **"uniform that we called his 'pest controller's outfit' ":** Lummitsch, "Erinnerungen," HC 1480.

164 **"paralysed and then meets with a lingering and agonising death":** Quoted in Keech, *Battleground Europe: St. Julien*, p. 36.

164 **"Urge that immediate steps be taken to supply similar means":** Quoted in L. F. Haber, *The Poisonous Cloud*, p. 41.

164 **aid the work of the army's Chemical Warfare Service:** Jones, "Chemical Warfare Research During World War I: A Model of Cooperative Research," in James Parascandola and James C. Whorton, eds., *Chemistry and Modern Society: Historical Essays in Honor of Aaron J. Ihde*, p. 166.

165 **"Haber traveled to the eastern front, where he was expected":** Willstätter, *Aus meinem Leben*, pp. 251–52.

166 **"agonized over his guilt in her suicide":** Jaenicke notes from conversation with Franck, April 16, 1958, HC 1449.

166 **"perversion of science":** Goran, *Story of Fritz Haber*, p. 71.

166 **"bonds remained with Fritz even after Clara's death":** Krassa to Jaenicke, November 2, 1957, HC 1470.

167 **detailed and lurid story, irresistible in its drama:** Lütke to Jaenicke, January 9, 1958, and January 17, 1958, HC 260.

167 **Hermann was the first to arrive at his mother's side:** Meyer to Jaenicke, November 9, 1958, HC 1483. Also Charlotte Haber, *Mein Leben mit Fritz Haber*, p. 90.

168 **"It was a time in which a human life meant little":** Willstätter, *Aus meinem Leben*, p. 232.

169 **"from between orders and telegrams, and I suffer":** Haber to Engler, June 12, 1915, HC 856. Quoted in Stoltzenberg, *Fritz Haber*, p. 356.

169 **"they take whatever they want, not what they need":** Quoted in Stoltzenberg, *Fritz Haber*, p. 255.

170 **Haber called it a "fabulous success":** Transcript of a meeting with representatives of German chemical industry, May 15, 1918, HC 522.

171 **"It is impossible for me to give a real idea of the terror and horror":** Quoted in Keech, *Battleground Europe: St. Julien*, pp. 58–59.

171 **"Gas! GAS! Quick boys!—An ecstasy of fumbling":** Owen, "Dulce et Decorum Est," in *Penguin Book of First World War Poetry*, pp. 192–93.

172 **"a higher form of killing":** This phrase is attributed to Haber in Harris and Paxman, *A Higher Form of Killing: The Secret History of Chemical and Biological Warfare*, front matter and p. 254.

173 **"Gas weapons and gas defense turn warfare into chess":** Haber to Duisberg, February 26, 1919, HC 860.

173 **"into a heap of helpless people":** Haber, *Fünf Vorträge*, p. 36.

173 **a new psychic challenge to the foe, "unsettling the soul":** Ibid., p. 37.

173 **"more fright and less destruction!":** Quoted in Szöllössi-Janze, *Fritz Haber*, p. 476.

174 **"The disapproval that the knight felt for the man with a gun":** Haber, *Fünf Vorträge*, p. 34.

174 **trenches in turn brought forth gas:** Haber, untitled memorandum, September 18, 1917, HC 1678.

175 **the harvest was expected to fall by 30 percent:** Minutes of meeting of nitrate commission on November 30, 1914, Archives of the Max Planck Society, X/12: NL Fischer, Film 2.

176 **would enrich Haber personally:** Szöllössi-Janze, *Fritz Haber*, pp. 185–91, 481–87.

177 **mediator in its acrimonious negotiations with the BASF:** Ibid., p. 302.

177 **the equivalent to about $4 million today:** John J. McCusker, "Comparing the Purchasing Power of Money in the United States (or Colonies) from 1665 to Any Other Year Including the Present," Economic History Services, 2004 (URL: http://www.eh.net/hmit/ppowerusd/), and Lawrence H. Officer, "Exchange Rate Between the United States Dollar and Forty Other Countries, 1913–1999," Economic History Services, EH.Net, 2002 (URL: http://www.eh.net/hmit/exchangerates/).

177 (about $1.4 billion in 2004 dollars) by the end of the war: Brock, "The Reactions of War: Arthur A. Noyes and the Quest for Nitrogen Compounds in the First World War," unpublished manuscript.

177 predicted that it would remain even after the war had ended: Dewey, "What Are We Fighting For?" p. 99.

178 "I received three proposals of marriage": Charlotte Haber, *Mein Leben mit Fritz Haber*, pp. 58–59.

179 "we belonged together, with soul and body": Ibid., p. 90.

179 "whether we possess that harmony": Quoted in Szöllössi-Janze, *Fritz Haber*, pp. 402–3.

179 "Miss Nathan's last journey": Haber to Charlotte Nathan, September 14, 1917, HC, Dep. L. Haber, 26.

180 disregard for her own desires: Quoted in Szöllössi-Janze, *Fritz Haber*, p. 405.

180 "thumbscrews of the soul": Quoted in ibid.

181 "There certainly are curious people in the world": Hartley, "Fritz Haber," undated, HC 1209.

182 but not for resource-starved Germany: Westphal, "68 Jahre als Physiker in Berlin," *Physikalische Blätter*, vol. 28 (1972), pp. 261–62.

182 "enough submarines *and* enough aircraft": Lummitsch to Jaenicke, July–August 1955, HC 1480.

TEN | LIKE FIRE IN THE HANDS OF CHILDREN

187 "Only scientific progress can restore all that the war destroyed": Haber, "Vorrede," *Fünf Vorträge*, n.p.

187 "like fire in the hands of small children": Haber to Kerschbaum, January 2, 1932, HC 1084.

188 "when you're on a snow-covered slope, sliding downward?": Haber to Duisberg, February 26, 1919, HC 860.

188 "until someone accidentally breaks it": Haber to Willstätter, January 7, 1926, in Werner and Irmscher, *Briefe*, p. 101.

189 "what I *can't* do, I no longer can learn": Ibid., p. 102.

189 scientists at his institute discovered a forged passport: Jaenicke notes from conversation with James Franck, April 16, 1958.

189 letters from an old acquaintance, the chemist Hermann Staudinger: The entire exchange of letters is held by the Deutsches Museum, Archiv, NL 088/DII 11.4.a-d. It is quoted extensively in Stoltzenberg, *Fritz Haber*, pp. 317–19.

191 passed the warning along to officials of the International Red Cross: Quoted in Stoltzenberg, *Fritz Haber*, p. 316.

193 **"It was a great experience to have enjoyed his confidence":** Hartley, "Fritz Haber," undated, HC 1209.

193 **its job apparently was to test potential new chemical weapons:** Szöllössi-Janze, *Fritz Haber,* pp. 464–66.

193 **the activities of Hugo Stoltzenberg:** These are described most comprehensively in Szöllössi-Janze, *Fritz Haber,* pp. 471–79.

195 **"I always jumped from one thing to another":** Haber to Willstätter, January 29, 1919, in Werner and Irmscher, *Briefe,* p. 56.

195 **a sense for what might lie beyond the horizon:** For insightful comments on Haber's intelligence, see Jaenicke's notes of a conversation with Max Meyer, November 9, 1958, HC 1483.

196 **"may lead to renewed international understanding":** Haber to Willstätter, November 18, 1919, in Werner and Irmscher, *Briefe,* p. 62.

196 **rejected their prizes in protest:** "Nobel Prizes That Go Begging," *New York Times,* January 27, 1920, p. 14.

196 **"who wrote General Ludendorff's daily communiqués":** Ibid.

196 **He wished to have no contact with Haber or Planck:** Crawford, *Nationalism and Internationalism in Science,* p. 67.

197 **might render every ammonia factory on earth obsolete:** The lecture can be found online at the Nobel Foundation website: www.nobel.se.

197 **"the government is helpless!":** Haber to Willstätter, 1919, in Werner and Irmscher, *Briefe,* p. 63.

201 **"I feel his character now that he is leaving us":** Quoted in Stoltzenberg, *Fritz Haber,* p. 13.

201 **the most reflective speech of his life:** The speech is reprinted in Haber, *Aus Leben und Beruf,* pp. 35–36.

203 **"It's nighttime, and I'm afraid to sleep":** Haber to Willstätter, August 21, 1927, in Werner and Irmscher, *Briefe,* pp. 110–11.

203 **"You asked how I'm doing. I'm suffering":** Haber to Willstätter, May 7, 1921, in Werner and Irmscher, *Briefe,* pp. 72–73.

204 **"In those hours I experience life with gratitude":** Haber to Willstätter, May 24, 1921, in Werner and Irmscher, *Briefe,* p. 75.

204 **"my damaged nerves and my diminished strength":** Haber to Willstätter, December 1924, in Werner and Irmscher, *Briefe,* p. 97.

205 **"forgetting the dire facts of politics":** Stern, "Fritz Haber: Personal Recollections," p. 94.

205 **"we got to hear a very clear and informative lecture":** Meitner to Jaenicke, December 4, 1954, HC 1484.

206 **"unexpected and always original and stimulating":** Coates, "The Haber Memorial Lecture," p. 1664.

206 **"He simply could react incredibly quickly":** Jaenicke notes from conversation with James Franck, April 16, 1958, HC 1449.

206 **"he would nod his head and say gravely, *'Aber! Aber! Aber!'*":** "Transcript of Interview with Hubert Alyea, May 22, 1986," p. 76, Oral History Collections, Archive of the Beckman Center for the History of Chemistry.

206 **through the garden in the direction of their laboratory:** Mueller to Jaenicke, August 23, 1958, HC 1488.

206 **"good nature was limitless when it came to such things":** Lummitsch, "Erinnerungen," HC 1480.

207 **gardener and its two mechanics, ten children in all:** Mueller to Jaenicke, August 23, 1958, HC 1488.

207 **"both your best friend and God at the same time":** Quoted in Sime, *Lise Meitner,* p. 142.

207 **"the same way as to another assistant":** Johannes Jaenicke notes from conversation with James Franck, April 16, 1958, HC 1449.

207 **which she did happily:** Meitner to Jaenicke, December 4, 1954, HC 1484.

208 **"others always want something from him":** Haber to Willstätter, June 16, 1923, in Werner and Irmscher, *Briefe,* p. 84.

208 **or even leaving Berlin altogether:** See Haber to Willstätter, May 13, 1921, and January 7, 1926, in Werner and Irmscher, *Briefe,* pp. 74, 100–102.

208 **"The thunderclap will echo around the world":** Haber to Willstätter, June 1924, in Werner and Irmscher, *Briefe,* p. 96.

208 **talked for an hour or more every day on the telephone:** Willstätter, *Aus meinem Leben,* p. 355.

209 **It would damage the reputation of German Jews for years to come:** Haber to Einstein, March 9, 1921, HC 978.

210 **"will try to visit you before his departure":** Einstein to Haber, March 9, 1921, HC 978.

210 **"of whom you are one of the most outstanding and benevolent":** Ibid.

210 **"Returning to our home in Dahlem was particularly hard":** Charlotte Haber, *Mein Leben mit Fritz Haber,* p. 269.

212 **Marga was subject to depression:** Ibid., pp. 120, 252.

213 **Stern remembered the trip as an idyllic time:** Stern, "Fritz Haber: Personal Recollections," pp. 86–89.

213 **"I wasn't able to get over the bitterness":** Charlotte Haber, *Mein Leben mit Fritz Haber,* p. 255.

213 **"Let us call ten years enough. I can't do it anymore":** Quoted in Szöllössi-Janze, *Fritz Haber,* p. 607.

213 **"necessary, if I was going to keep on living"**: Haber to Einstein, December 29, 1928, quoted in Stoltzenberg, *Fritz Haber*, p. 395.

214 **"I don't want to stay here"**: Haber to Willstätter, December 26, 1927, in Werner and Irmscher, *Briefe*, p. 112.

215 **"fire in the hands of small children"**: Haber to Kerschbaum, January 2, 1932, HC 1084.

216 **"I've made serious mistakes in life"**: Haber to Willstätter, February 24, 1933, in Werner and Irmscher, *Briefe*, p. 123.

ELEVEN | DISPOSSESSION

217 **"to the ammonia-breathing German war god"**: Roth, "The Auto-da-Fé of the Mind" (1933), in Roth, *What I Saw: Reports from Berlin, 1920–1933*, p. 209.

219 **"even worse than in 1918"**: Haber to Dietrich, May 18, 1931, HC 955.

219 **"selfishness leads to the greatest well-being of the whole"**: Haber to Duisberg, May 31, 1918, HC 859.

219 **"Back then, I felt that it was too early"**: Haber to Dietrich, May 18, 1931, HC 955.

219 **"force of public sentiment behind it"**: Ibid.

220 **in campaign funding from sympathetic industrialists:** Haber to Hermann Haber, February 1933, HC 1036.

221 **he and Willstätter now sat in the same wildly rocking boat:** Haber to Willstätter, April 1, 1933, in Werner and Irmscher, *Briefe*, pp. 127–28.

222 **"don't scold your James Franck, who loves you"**: Franck to Haber, April 15, 1933, HC 991.

222 **"all non-Aryan as defined by the law"**: Quoted in Szöllössi-Janze, *Fritz Haber*, p. 647.

223 **"now dictates this request for retirement"**: Quoted in Willstätter, *Aus meinem Leben*, p. 273.

224 **"I'm finished with the Jew Haber"**: Stern, "Fritz Haber: Personal Recollections," p. 99.

224 **Planck could only leave the room:** Planck, "Mein Besuch bei Adolf Hitler," *Physikalische Blätter*, vol. 3 (1947), p. 143.

224 **That myth was reborn in Nazi propaganda:** Mommsen, *Imperial Germany*, p. 215.

224 **"I was German to an extent that I feel fully only now"**: Haber to Willstätter, after April 1933, in Werner and Irmscher, *Briefe*, p. 132.

225 **"his ancestors and he himself served to the best of their ability"**: Haber to Donnan, date uncertain (the letter carries the date March 24, 1933, but was almost certainly written later, perhaps in May), HC 958.

226 **"I never believed in it in the least":** Einstein to Haber, May 19, 1933, HC 983. Quoted in Fritz Stern, *Einstein's German World,* p. 159.

226 **"I made a feeble attempt to comfort him":** Weizmann, *Trial and Error,* p. 353.

227 **"I knew the battlefields on which French and English Jews":** Haber to Weizmann, January 6, 1934, in "Letters to Chaim Weizmann," Yearbook of the Leo Baeck Institute, 1963, p. 110.

227 **"The old man is completely kaputt":** Quoted in Stoltzenberg, *Fritz Haber,* p. 599.

228 **"with honor but without heavy duties":** Haber to Pope, August 4, 1933. HC 2321.

228 **branded in his homeland a tax evader:** Hermann Haber to Weizmann, July 26, 1933, HC 823.

229 **"as an act of international courtesy, to support it":** Haber to Pope, August 4, 1933, HC 2321.

229 **"I was never in my life so Jewish as now!":** Haber to Einstein, undated, HC 983.

230 **"a few fine persons (Planck 60% noble and [Max von] Laue 100%)":** Einstein to Haber, August 9, 1933, HC 983.

230 **"My poor friend took medications":** Willstätter, *Aus meinem Leben,* p. 390.

231 **"this is absolutely necessary in these times":** Haber to Ludwig Haber, August 17, 1933, HC 1056.

231 **"I will accept your generous invitation":** Haber to Pope, August 23, 1933, HC 2321.

231 **Weizmann recalls Haber breaking into an "eloquent tirade":** Weizmann, *Trial and Error,* p. 354.

232 **"With these words I depart":** "Habers Abschieds-Kundgebung," October 1, 1933, HC 618.

233 **"made worse by circumstances that no one can change":** Hermann Haber to Weizmann, October 29, 1933, HC 823.

233 **"leadership functions that I built up over 40 years at home":** Haber to Weizmann, January 15, 1934, in "Letters to Chaim Weizmann," p. 111.

234 **He climbed a "skyscraper of hopes":** Haber to Weizmann, January 6, 1934, in "Letters to Chaim Weizmann," p. 109.

235 **"a hundred years without physics and chemistry!":** Willstätter, *Aus meinem Leben,* p. 274.

235 **"You are now a man of decisive importance":** Haber to Bosch, undated, but probably written in December 1933 HC 944.

235 **that gasoline would fuel Hitler's blitzkrieg:** The contract was an echo of the BASF's contracts to deliver ammonia to the German military during World War I. It would be introduced at the war crimes trials of Nuremberg as evidence of a conspiracy between I. G. Farben and Nazi rulers aimed at preparing for wars of aggression. The judges, however, rejected the charges as unproven. See Hughes, "Technological Momentum in History: Hydrogenation in Germany, 1898–1933," *Past and Present*, vol. 44 (1969), pp. 106–32.

236 **"This is my last night in your country":** Haber to Pope, January 25, 1934, HC 2321.

237 **"discussed plans for the future":** Stern, "Fritz Haber: Personal Recollections," p. 102.

238 **"a servant of his homeland":** Hermann Haber to unknown, undated, HC 1375.

TWELVE | REQUIEM

239 **"they can't really be expressed in words":** Haber to Willstätter, January 22, 1934, in Werner and Irmscher, *Briefe*, p. 142.

241 **"the tragedy of unrequited love":** Einstein to Hermann and Marga Haber, undated, presumably 1934, HC 58.

241 **"service of his nation and all of humanity":** Laue, "Fritz Haber," in *Die Naturwissenschaften*, vol. 22 (1943), p. 97.

242 **prohibiting all civil servants, including university faculty, from attending the event:** Kunisch to rectors of German universities, January 15, 1935, Historisches Archiv Krupp, Essen, FAH 4E251. This document is attached to a lengthy confidential report (hereafter, Planck report) that Planck wrote after the event was over. Historian Kristie Macrakis discovered it in the mid-1990s.

244 **more vividly than the ones before him:** Willstätter, *Aus meinem Leben*, p. 276.

244 **"not to outside observers, and not to Dahlem":** Hermann Haber to Planck, undated, HC Dep. L. Haber 1375.

244 **who'd come at his telegraphed request:** Hahn, "Zur Erinnerung an die Haber-Gedächtnisfeier," *Mitteilungen aus der Max-Planck-Gesellschaft*, vol. 1 (1960), p. 12.

245 **who'd been ordered to stay away:** Ibid., p. 13.

245 **might yet appear to disrupt the meeting:** Planck report, February 6, 1935, Historisches Archiv Krupp, Essen, FAH 4E251.

245 **a haunting string quartet by Franz Schubert:** Hahn, "Zur Erinnerung an die Haber-Gedächtnisfeier," p. 6. The service began with the *Andante con moto* movement from Schubert's Quartet No. 14.

245 **"a fighter for Germany"**: Planck report, February 6, 1935, Historisches Archiv Krupp, Essen, FAH 4E251.

245 **"more than any other single individual to Germany's power"**: "Nazis Gag Haber Services," *New York Times,* January 26, 1935, p. 10. Also "Haber Memorial Is Held in Berlin," *New York Times,* January 30, 1935, p. 6.

245 **Haber had intended this poison to protect human life:** See Szöllössi-Janze, *Fritz Haber,* pp. 462–64.

THIRTEEN | THE HEIRS

247 **"production of asphyxiant gas in Germany during the last war"**: Stern, *Grandeurs et défaillances de l'Allemagne du xxᵉ siècle,* p. 308.

249 **"is essential to the modern state"**: L. F. Haber, *The Chemical Industry, 1900–1930,* p. 2.

250 **"The generation gap has its uses"**: L. F. Haber, *The Poisonous Cloud,* pp. 1–14.

252 **the loyal son of any good German must certainly be disloyal to France:** Stern, *Grandeurs,* pp. 308–11.

252 **"finally have gotten a real start here"**: Marga Haber to Eisner, December 22, 1937, HC 833. See also Hermann Haber to unknown, October 19, 1937, HC 827.

253 **three daughters arrived in Hoboken, New Jersey:** Stern, *Grandeurs,* pp. 313–14.

254 **"an inescapable moral and historical conundrum"**: Stern, *The Failure of Illiberalism,* p. xiv.

254 **"I could not abandon the subject"**: Stern, *Dreams and Delusions,* p. 26.

254 **to protect his father's friend against unfair attacks:** Ibid., pp. 51–76.

255 **"Haber an impresario of collective greatness"**: Stern, *Einstein's German World,* p. 62.

255 **"Death to the French"**: Deutsches Chemiemuseum Merseburg, *Museumsführer durch den Technikpark,* 2002, p. 11.

256 **what one historian later called "technological momentum"**: Hughes, "Technological Momentum in History," *Past and Present,* vol. 44 (1969), p. 114.

256 **Leuna delivered gasoline fuel for the German blitzkrieg:** Hughes, "Technological Momentum in History," *Past and Present* 44 (1969), pp. 106–32.

256 **who weren't allowed into bomb shelters:** Streller and Maßalsky, *Geschichte des VEB Leuna-Werke "Walter Ulbricht": 1916–1945,* pp. 198–99.

257 **enormous craters filled with lakes as acid as vinegar:** Charles, "Wasteworld," *New Scientist,* January 31, 1998, pp. 32–35.

FOURTEEN | LESSONS LEARNED

259 **"infinitely greater dangers than were all the inventions of the past"**: Smith, *A Peril and a Hope*, p. 372.

260 **could keep pace with its technical achievements:** Willstätter, *Aus meinem Leben*, p. 229.

261 **to a responsible official of the U.S. government:** Smith, *A Peril and a Hope*, p. 31.

263 **"love for the blond beast has cooled off a bit"**: Einstein to Haber, August 9, 1933, HC 983.

264 **"like fire in the hands of small children"**: Haber to Kerschbaum, January 2, 1932, HC 1084.

Bibliography

ARCHIVES

Archives for the History of the Max Planck Society, Berlin, Germany.
　Haber Collection (HC).
　Emil Fischer Papers (microfilm).
Archive of the Beckman Center for the History of Chemistry, Philadelphia.
　Oral history collections.
Library of Congress, Manuscript Collection, Washington, DC.
　Woodrow Wilson manuscripts.
Deutsches Museum, Munich, Germany.
　Hermann Staudinger Papers.
Historisches Archiv Krupp, Essen, Germany.

BOOKS, ARTICLES, AND OTHER LITERATURE

Adams, Henry. *The Education of Henry Adams.* 1907. Reprint, with introduction and notes by Ernest Samuels. Boston: Houghton Mifflin Company, 1973. Page references are to the 1973 edition.

Architecture Museum of Wroclaw. *Der alte jüdische Friedhof in Wroclaw.* Wroclaw, Poland: Architecture Museum of Wroclaw, 1988.

Blackbourn, David. *The Long Nineteenth Century: A History of Germany, 1780–1918.* New York: Oxford University Press, 1998.

Boehlich, Walter, ed. *Der Berliner Antisemitismusstreit.* Frankfurt/Main: Insel Verlag, 1965.

Brauch, Hans Günter, and Rolf-Dieter Müller, eds. *Chemische Kriegführung, chemische Abrüstung: Dokumente und Kommentare.* Berlin: A. Spitz, 1985.

Brock, David. "The Reactions of War: Arthur A. Noyes and the Quest for Nitrogen Compounds in the First World War." Unpublished paper.

Brock, William, H. *The Norton History of Chemistry.* New York: W. W. Norton, 1993.

Brocke, Bernhard vom. "Wissenschaft und Militarismus: Der Aufruf der 93 'An die Kulturwelt' und der Zusammenbruch der internationalen Gelehrtenrepublik im Ersten Weltkrieg." In *Wilamowitz nach 50 Jahren,* edited by William Calder II, Hellmut Flashar, and Theodor Lindken. Darmstadt: Wissenschaftliche Buchgesellschaft, 1985. Pp. 649–717.

Charles, Daniel. "Wasteworld," *New Scientist,* January 31, 1998: 32–35.

Coates, J. E. "The Haber Memorial Lecture." Delivered before the Chemical Society on April 29, 1937. Published in *Journal of the Chemical Society,* 1939, Part II: 1642–72.

Cohen, Ruth Schwartz. *A Social History of American Technology.* New York: Oxford University Press, 1997.

Crawford, Elisabeth. *Nationalism and Internationalism in Science, 1880–1939: Four Studies of the Nobel Population.* Cambridge, U.K.: Cambridge University Press, 1992.

Crookes, William. *The Wheat Problem.* 3d ed. London: Longmans, Green & Co., 1917.

Davies, Norman, and Roger Moorhouse. *Microcosm: Portrait of a Central European City.* London: Jonathan Cape, 2002.

Deutsches Chemiemuseum Merseburg. *Museumsführer durch den Technikpark.* Merseburg: Deutsches Chemiemuseum Merseburg, 2002.

Dewey, John. "What Are We Fighting For?" (1918). In *The Middle Works.* Vol. 11. Carbondale and Edwardsville: Southern Illinois University Press, 1982.

Elon, Amos. *The Pity of It All: A History of the Jews in Germany, 1743–1933.* New York: Metropolitan Books, 2002.

Feldman, Gerald. "Hugo Stinnes and the Prospect of War." In *Anticipating Total War,* edited by Manfred Boemeke, Roger Chickering, and Stig Förster. Cambridge, U.K.: Cambridge University Press, 1999. Pp. 92–93.

Galloway, James, and Ellis Cowling. "Reactive Nitrogen and the World: 200 Years of Change." *Ambio* 31, no. 2 (March 2002): 64–71.

Geyer, Michael. *Aufrüstung oder Sicherheit: Die Reichswehr in der Krise der Machtpolitik 1924–1936.* Wiesbaden: Franz Steiner Verlag, 1980.

Goolsby, Donald, William Battaglin, Brent Aulenbach, and Richard P. Hooper. "Nitrogen Input to the Gulf of Mexico." *Journal of Environmental Quality* 30 (2001): 329–36.

Goran, Morris. *The Story of Fritz Haber.* Norman: University of Oklahoma Press, 1967.

Haber, Charlotte. *Mein Leben mit Fritz Haber.* Düsseldorf: Econ Verlag, 1970.

Haber, Fritz. *Aus Leben und Beruf.* Berlin: Verlag von Julius Springer, 1927.

———. *Fünf Vorträge.* Berlin: Verlag von Julius Springer, 1924.

———. "Über Hochschulunterricht und elektrochemische Technik in den Vereinigten Staaten." *Zeitschrift für Elektrochemie* 9 (1903): 291–406.

———. "Vorträge." *Zeitschrift für Elektrochemie* 9 (1903): 894.

Haber, L. F. *The Chemical Industry, 1900–1930: International Growth and Technological Change.* Oxford, U.K.: Clarendon Press, 1971.

———. *The Poisonous Cloud: Chemical Warfare in the First World War.* Oxford, U.K.: Oxford University Press, 1986.

Hahn, Otto. *Mein Leben.* Munich: Bruckman, 1969.

———. "Zur Erinnerung an die Haber-Gedächtnisfeier." *Mitteilungen aus der Max-Planck-Gesellschaft* 1 (1960): 3–13.

Hamecher, Hirst. *Königin der See: Fünfmast-Vollschiff* Preussen. Hamburg: Verlag Chronik der Seefahrt Heinemann, 1969.

Hanslian, Rudolf. *Der deutsche Gasangriff bei Ypern am 22. April 1915: Eine kriegsgeschichtliche Studie.* Berlin: Verlag Gasschutz und Luftschutz, 1934.

Harnack, Adolf. *Aus Wissenschaft und Leben.* Giessen: A. Töpelmann, 1911.

Harris, Robert, and Jeremy Paxman. *A Higher Form of Killing: The Secret History of Chemical and Biological Warfare.* Random House trade paperback ed. New York: Random House, 2002.

Henao, Julio, and Carlos Baanante. "Nutrient Depletion in the Agricultural Soils of Africa." *2020 Brief No. 62.* Washington, D.C.: International Food Policy Research Institute, 1999.

Hoffman, Roald. *The Same and Not the Same.* New York: Columbia University Press, 1995.

———, and Pierre Laszlo. "Coping with Fritz Haber's Somber Literary Shadow." *Angewandte Chemie International Edition* 40 (2001): 4599–4604.

Holdermann, Karl. *Im Banne der Chemie: Carl Bosch, Leben und Werk.* Düsseldorf: Econ Verlag, 1953.

Hughes, Thomas P. *American Genesis: A Century of Invention and Technological Enthusiasm, 1870–1970.* New York: Viking Press, 1989.

———. "Technological Momentum in History: Hydrogenation in Germany, 1898–1933." *Past and Present* 44 (1969): 106–32.

Ipatieff, Vladimir N. *The Life of a Chemist.* Palo Alto, Calif.: Stanford University Press, 1946.

Jansen, Sarah. *"Schädlinge": Geschichte eines wissenschaftlichen und politischen Konstrukts 1840–1920.* Frankfurt: Campus, 2003.

Jarausch, Konrad. *The Enigmatic Chancellor: Bethmann-Hollweg and the Hubris of Imperial Germany, 1856–1921.* New Haven, Conn.: Yale University Press, 1973.

———. *Students, Society, and Politics in Imperial Germany: The Rise of Academic Illiberalism.* Princeton, N.J.: Princeton University Press, 1982.

————. "The Universities: An American View." In *Another Germany: A Reconsideration of the Imperial Era*, edited by Jack R. Dukes and Joachim Remak. Boulder, Colo.: Westview Press, 1988. Pp. 181–206.

Johnson, Jeffrey. *The Kaiser's Chemists: Science and Modernization in Imperial Germany.* Chapel Hill: University of North Carolina Press, 1990.

Jones, Daniel P. "Chemical Warfare Research During World War I. A Model of Cooperative Research." In *Chemistry and Modern Society: Historical Essays in Honor of Aaron J. Ihde*, edited by John Parascandola and James C. Whorton. Washington, D.C.: American Chemical Society, 1983.

Kaiser-Wilhelm-Gesellschaft. *25 Jahre Kaiser-Wilhelm-Gesellschaft zur Förderung der Wissenschaften.* Berlin: Verlag von Julius Springer, 1936.

Keech, Graham. *Battleground Europe: St. Julien.* Barnsley, South Yorkshire: Pen & Sword Books, 2001.

Kevles, Daniel J. "George Ellery Hale, the First World War, and the Advancement of Science in America." *Isis* 59 (1968): 427–37.

Klemperer, Victor. *Curriculum Vitae: Jugend um 1900.* Berlin: Siedler, 1989.

Kolb, Eberhard. *The Weimar Republic.* London: Unwin Hyman, 1988.

Lagiewski, Maciej. *Wroclaw wczoraj—Breslau gestern.* Gliwice, Poland: Wydawnictwo "Wokol nas," 1998.

Landes, David. *The Unbound Prometheus: Technological Change and Industrial Development in Western Europe from 1750 to the Present.* Cambridge, U.K.: Cambridge University Press, 1969.

Lanouette, William, with Bela Szilard. *Genius in the Shadows: A Biography of Leo Szilard, the Man Behind the Bomb.* New York: Charles Scribner's Sons, 1992.

Laue, Max von. "Fritz Haber." *Naturwissenschaften* 22 (1934): 97.

Le Rossignol, Robert. "Zur Geschichte der Herstellung des synthetischen Ammoniaks." *Die Naturwissenschaften* 16 (1928): 1070–71.

Leigh, G. J. *The World's Greatest Fix: A History of Nitrogen and Agriculture.* New York: Oxford University Press, 2004.

Leitner, Gerit von. *Der Fall Clara Immerwahr: Leben für eine humane Wissenschaft.* 2d ed. Munich: C. H. Beck, 1994.

Macdonald, Lyn. *1915: The Death of Innocence.* London: Headline, 1993.

MacDonogh, Giles. *Berlin.* New York: St. Martin's Press, 1998.

McMurray, Jonathan S. *Distant Ties: Germany, the Ottoman Empire, and the Construction of the Baghdad Railway.* Westport, Conn.: Praeger, 2001.

Macrakis, Kristie. *Surviving the Swastika: Scientific Research in Nazi Germany.* New York: Oxford University Press, 1993.

Maier, Pauline, Merritt Roe Smith, Alexander Keyssar, and Daniel J. Kevles. *Inventing America: A History of the United States.* New York: W. W. Norton, 2003.

Mann, Golo. *The History of Germany Since 1789.* London: Chatto & Windus, 1972.

Mann, Thomas. *Doctor Faustus.* Translated by John E. Wood. New York: Knopf, 1997.

———. *The Magic Mountain.* Translated by H. T. Lowe-Porter. London: M. Secker, 1927.

———. *Reflections of a Nonpolitical Man.* Translated by Walter D. Morris. New York: Unger, 1983.

Martinetz, Dieter. *Der Gaskrieg 1914–18: Entwicklung, Herstellung und Einsatz chemischer Kampfstoffe.* Bonn: Bernard & Graefe Verlag, 1996.

Max-Planck-Gesellschaft zur Förderung der Wissenschaften. *50 Jahre Kaiser-Wilhelm-Gesellschaft und Max-Planck-Gesellschaft zur Förderung der Wissenschaften, 1911–1961: Beiträge und Dokumente.* Göttingen: Max-Planck-Gesellschaft, 1961.

Mitchell, E. J. (pseudonym). "Trip Report: Peking, China." *Chemical Engineering,* July 9, 1973, pp. 92–98.

Mommsen, Wolfgang. *Bürgerstolz und Weltmachtstreben: Deutschland unter Wilhelm II.* Berlin: Propyläen Verlag, 1995.

———. *Imperial Germany, 1867–1918: Politics, Culture, and Society in an Authoritarian State.* Translated by Richard Deveson. London: Arnold, 1995.

———. "The Topos of Inevitable War in Germany in the Decade Before 1914." In *Germany in the Age of Total War,* edited by Volker Berghahn and Martin Kitchen. London: Croom Helm, 1981. Pp. 23–45.

Nietzsche, Friedrich. *Untimely Meditations.* Edited by Daniel Breazeale. Translated by R. J. Hollingdale. Cambridge, U.K.: Cambridge University Press, 1997.

Ostwald, Wilhelm. *Lebenslinien: Eine Selbstbiographie.* Berlin: Clasing & Co., 1926–27.

Owen, Wilfred. "Dulce et Decorum Est." In *The Penguin Book of First World War Poetry,* edited by Jon Silkin. 2d ed. London: Penguin, 1996. Pp. 192–93.

Planck, Max. "Mein Besuch bei Adolf Hitler." *Physikalische Blätter* 3 (1947): 143.

Plumpe, Gottfried. *Die I. G. Farbenindustrie AG: Wirtschaft, Technik und Politik 1904–1945.* Berlin: Duncker & Humblot, 1990.

Radkau, Joachim. *Das Zeitalter der Nervosität: Deutschland zwischen Bismarck und Hitler.* Munich: Propyläen Taschenbuch, 2000.

Reid, Constance. *Hilbert.* With an appreciation of Hilbert's mathematical work by Hermann Weyl. New York: Springer-Verlag, 1970.

Rhodes, Richard. *The Making of the Atomic Bomb.* New York: Simon & Schuster, 1986.

Rosegrant, Mark W., Michael S. Paisner, Siet Meijer, and Julie Witcover. *Global Food Projections to 2020: Emerging Trends and Alternative Futures.* Washington, D.C.: International Food Policy Research Institute, 2001.

Roth, Joseph. *What I Saw: Reports from Berlin, 1920–1933.* Translated by Michael Hofmann. New York: W. W. Norton, 2003.

Sime, Ruth Lewin. *Lise Meitner: A Life in Physics.* Berkeley: University of California Press, 1996.

Skaggs, Jimmy M. *The Great Guano Rush: Entrepreneurs and American Overseas Expansion*. New York: St. Martin's Press, 1994.

Smil, Vaclav. *Enriching the Earth: Fritz Haber, Carl Bosch, and the Transformation of World Food Production*. Cambridge, Mass.: MIT Press, 2001.

Smith, Alice Kimball. *A Peril and a Hope: The Scientists' Movement in America, 1945–47*. Cambridge, Mass.: MIT Press, 1970.

Stern, Fritz. *Dreams and Delusions: The Drama of German History*. New York: Knopf, 1987.

————. *Einstein's German World*. Princeton, N.J.: Princeton University Press, 1999.

————. *The Failure of Illiberalism: Essays on the Political Culture of Modern Germany*. New York: Columbia University Press/Morningside, 1992.

————. *Five Germanies I Have Known*. Wassenaar, The Netherlands: Netherlands Institute for Advanced Study, 1998.

————. *Grandeurs et défaillances de l'Allemagne du xxᵉ siècle: Le cas exemplaire d'Albert Einstein*. Translated by Michel Charlot, Pierre Duchamp, and Denis Trierweiler. Paris: Fayard, 2001.

Stern, Rudolf. "Fritz Haber: Personal Recollections." In *Yearbook of the Leo Baeck Institute* (1963). New York: Leo Baeck Institute, 1963. Pp. 70–102.

Stoltzenberg, Dietrich. *Fritz Haber: Chemiker, Nobelpreisträger, Deutscher, Jude*. Weinheim: Wiley-VCH, 1998.

————. *Fritz Haber: Chemist, Nobel Laureate, German, Jew*. Philadelphia: Chemical Heritage Foundation, 2004.

Streller, Karl-Heinz, and Erika Maßalsky. *Geschichte des VEB Leuna-Werke "Walter Ulbricht": 1916–1945*. Leipzig: VEB Deutscher Verlag für Gundstoffindustrie, 1989.

Strupp, Christoph. "War in Academe: American Perceptions of German Science and Research During World War I." Paper presented at conference organized by the Institute for the History of Science and Technology, Russian Academy of Science, in St. Petersburg, Russia, April 8–10, 2003.

Szöllössi-Janze, Margit. *Fritz Haber 1868–1934: Eine Biographie*. Munich: C. H. Beck, 1998.

Thiessen, Vern. *Einstein's Gift*. Toronto: Playwright's Canada Press, 2003.

Trescott, Martha Moore. *The Rise of the American Electrochemicals Industry, 1880–1910: Studies in the American Technological Environment*. Westport, Conn.: Greenwood Press, 1981.

Trumpener, Ulrich. "The Road to Ypres." *Journal of Modern History* 47 (September 1975): 460–80.

Verhey, Jeffrey. *The Spirit of 1914: Militarism, Myth and Mobilization in Germany*. Cambridge, U.K.: Cambridge University Press, 2000.

Vierhaus, Rudolf, and Bernhard vom Brocke, eds. *Forschung im Spannungsfeld von Politik und Gesellschaft: Geschichte und Struktur der Kaiser-Wilhelm-/Max-Planck-Gesellschaft*. Stuttgart: Deutsche Verlags-Anstalt, 1990.

Von Hippel, Frank, and Joel Primack. *Advice and Dissent*. New York: Basic Books, 1974.

Weizmann, Chaim. *Trial and Error*. New York: Harper, 1949.

Werner, Petra, and Angelika Irmscher, eds. *Fritz Haber: Briefe an Richard Willstätter, 1910–1934*. Berlin: Verlag für Wissenschafts- und Regionalgeschichte Dr. Michael Engel, 1995.

Westphal, Wilhelm Heinrich. "68 Jahre als Physiker in Berlin." *Physikalische Blätter* 28 (1972): 258–65.

Whitman, Walt. "Passage to India" (1868). In Malcolm Cowley, ed., *The Complete Poetry and Prose of Walt Whitman*. Garden City, N.Y.: Garden City Books, 1948.

Wille, Hermann Heinz. *Der Januskopf: Leben und Wirken des Physikochemikers und Nobelpreisträgers Fritz Haber*. Berlin: Verlag Neues Leben, 1969.

Willstätter, Richard. *Aus meinem Leben*. Weinheim: Verlag Chemie, 1949.

Zeidler, Manfred. *Reichswehr und Rote Armee, 1920–1933: Wege und Stationen einer ungewöhnlichen Zusammenarbeit*. Munich: R. Oldenbourg Verlag, 1993.

Acknowledgments

HAD IT NOT BEEN for Fritz Haber's friends, the details of his life would quickly have disappeared and no biography would have been possible. Haber died an outcast, without a permanent address or a professional home to safeguard his files. The cataclysms of war and genocide scattered his family and most of his closest friends across two continents. Correspondence lay abandoned in basements and bombs destroyed some of the archives containing records of Haber's work.

Several of Haber's friends and collaborators—J. E. Coates, Richard Willstätter, and Rudolf Stern—published short accounts of Haber's life before they died. One of Haber's young students, Johannes Jaenicke, attempted a full-scale biography. He spent thirty years, beginning in 1954, combing through archives and tracking down Haber's friends and family members, interviewing them and making copies of letters and documents that survived. As a writer, however, Jaenicke was insurmountably blocked. At the end of the 1980s, he admitted defeat and delivered his files to the Archives for the History of the Max Planck Society for the Advancement of Science in Berlin, which in turn made them available to researchers.

The Jaenicke files became the documentary foundation of this book, and if there's an archive that offers more pleasant working conditions and a more helpful staff than the Archives for the History of the Max Planck Society, I have never encountered it. I'm grateful to Eckart Henning, the archive's director, along with Dirk Ullmann, Susanne Uebele, and Bernd Hoffman, for their unfailing and cheerful aid.

I also benefited enormously from the work of Margit Szöllössi-Janze, at the University of Cologne, who published a monumental biography of Haber, *Fritz Haber 1868–1934: Eine Biographie,* in 1998. Szöllössi-Janze's book is available only in German, but for anyone looking to explore Haber's life and times in greater detail, this is the place to go. Professor Szöllössi-Janze also took the time at various points to answer my questions. I'm grateful as well to an old friend, Johannes Toaspern, who in 1989 introduced me to the Leuna-Werke, a monstrous industrial ruin near Halle, Germany, that represented the peak of Fritz Haber's power and wealth when it was built nearly a century ago.

Before I had any reasonable prospects for writing a book on Fritz Haber, Alison Richards, a science editor at National Public Radio, gave me a chance to explore Haber's life and legacy in a pair of radio reports. I'm grateful to her for supporting this unconventional project. The reports were made possible through a grant from the Alfred P. Sloan Foundation.

The door to the publishing world did open, thanks to the persuasive efforts of my agent, Katinka Matson. And this book was fortunate to fall into the capable hands of Dan Halpern and his delightful crew at Ecco Press. My thanks go to everyone there—Gheña Glijansky, Mareike Paessler, Jill Bernstein, Amy Baker, and others whom I haven't yet met—but especially to Julia Serebrinsky, who edited this book. No writer could hope for a wiser, sharper, or more cheerful editor.

Many institutions and individuals aided my research. They include the Historisches Archiv Krupp, the Deutsches Museum, the Leo Baeck Institute, Matthias Scheffler at the Fritz Haber Institute of the Max Planck Society, Christoph Strupp, David Brock at the Chemical Heritage Foundation, Eva Lewis, Fritz Stern, Julian Perry Robinson, Frank von Hippel, Les Southerland at CF Industries' Donaldsonville Plant, and Manfred Mayer at the Technical University of Karlsruhe. I am grateful to Carol Smucker for providing expert translations of French texts, and to Elisabeth Mait and Christoph Strupp, from the German Historical Institute in Washington, D.C., who deciphered the Sütterlin script in a number of handwritten letters.

I'll never be able to adequately thank all the friends who spent precious free time reading this manuscript and suggesting ways to make it better. Noel Gunther and the Palo Alto–based team of Eric Weiner and Sharon Moshavi became full-fledged (though unpaid) editors, devoting countless hours to the task. My heartfelt thanks for gracious and generous editorial advice also go to Margery Hall, David Rosenthal, Carolyn Marzke, Peter Pringle, Alison Richards, Nancy Watzman, Christopher Joyce, Rosanna Landis, John Hiebert, and David Kestenbaum. Dolores Augustine, from St. John's University, rescued me from errors concerning German history. James Galloway, from the University of Virginia, and Keith Wiebe, from the U.S. Department of Agriculture's Economic Research Service, provided similar help for the sections dealing with nitrogen in agriculture. None of them should be blamed, of course, for any errors that remain.

I owe the greatest debt of appreciation and gratitude, however, to the three people who have shared the joys, hopes, and anxieties of this venture over the last two years, who've made room in their lives for my moods and my mess: Brigid, my peerless wife, and our children, Molly and Nora. Brigid first pointed me toward Fritz Haber's story four years ago, and her support and encouragement, always clearheaded and unsentimental, saw it through to the end. On some occasions, I wrote while she played violin or piano. I hope that the words of this book reflect, in some small measure, the spirit and devotion that I heard in her music. And to Molly and Nora, I say thanks for your laughter, your drawings, and your own stories. This particular tale may not be one you'll choose to read for a while. But in the meantime, I'll tell you a few more about our old friend Fiddlestick, the mouse.

Index